The International Space Station Zero Gravity, Maximum Discovery

© 2024 by Dr. Leo Lexicon

Notice of Rights

All rights reserved. No part of this publication may be reproduced, distributed, or transmitted in any form by any means, including photocopying, recording, or any information storage and retrieval systems, without the prior written permission of the author, except in the case of brief quotations embodied in critical reviews and certain other noncommercial uses permitted by copyright law. For permissions, please contact the author.

Copyright Infringement Disclaimer

Reproduction or translation of any part of this work beyond the extent permitted by Sections 107 or 108 of the 1976 United States Copyright Act, and any companion works, summaries, or guides, is forbidden without authorization.

Liability Disclaimer

The content provided herein is for informational purposes only, with no implied warranties of accuracy, relevance, or timeliness. The author and publisher disclaim all liability for any personal loss or damage arising indirectly or directly from the use of this material, which is provided "as is," and readers assume full responsibility for their use of the information. The cover image is AI-generated and not a direct representation of any person, living or dead. It is an artistic interpretation for illustrative purposes only, with no endorsements or affiliations implied. Liability related to image resemblance to real persons is disclaimed.

This publication may contain affiliate links, disclosed in accordance with the Federal Trade Commission guidelines. The author appreciates the support of readers who use these links, contributing to the content's ongoing creation.

All references and links provided in this book are for informational purposes only. While every effort has been made to provide the most accurate and up-to-date information, no warranty is given regarding their accuracy or content. The reader bears sole responsibility for the use of these references and agrees to use the information at their own risk.

The International Space Station

Zero Gravity, Maximum Discovery

by

Dr. Leo Lexicon

The International Space Station

Zero Gravity, Maximum Discovery

Embark on an extraordinary journey through the creation, life, and legacy of the International Space Station (ISS)—a marvel of engineering and a symbol of peaceful international collaboration. This book provides an in-depth look at the challenges, breakthroughs, and human stories that shaped one of the most significant scientific achievements in history.

From the early concepts of space stations and the Cold War collaborations that eventually gave birth to this unparalleled space structure, to the rigorous engineering feats and the unique day-to-day life of astronauts aboard, the story of the ISS is one of perseverance, innovation, and diplomacy. As you move through each chapter, you will explore not only the technical triumphs but also the cultural and psychological dynamics of life in space, showcasing the ISS as a microcosm of global cooperation.

What You Will Learn:

1. **Dawn of the ISS Vision** – Discover the origins of the ISS from early space station concepts to the Cold War-era collaborations that set the stage for international partnership.
2. **Engineering the Impossible** – Learn how engineers overcame technical challenges to design, build, and assemble the ISS in one of the most hostile environments imaginable—space.

3. **Assembly in Orbit** – Follow the complex construction process as astronauts became builders in space, overcoming setbacks and achieving milestones to complete the station's framework.
4. **Life on the ISS** – Get an inside look at the daily life of astronauts, their groundbreaking scientific experiments, and the unique cultural experiences they share while living in zero gravity.
5. **Scientific Contributions** – Familiarize yourself with the scientific achievements of the ISS, from medical research to Earth and space observations, and how international research collaborations have advanced human knowledge.
6. **Challenges and Crises** – Examine the technical, political, and logistical challenges faced during the ISS's operations, including space hazards and crisis management aboard.
7. **Legacy and Future** – Reflect on the ISS's role as a symbol of peaceful cooperation, its contributions to deep space exploration, and its lasting impact on future generations of scientists and engineers.

Why the International Space Station Matters:

As the world's largest and most complex space structure, the ISS stands as a testament to what humanity can achieve when we work together across national borders. This book not only chronicles the historical significance of the ISS but also looks ahead to its legacy—how the knowledge and technologies developed aboard the ISS are shaping the future of deep space exploration and inspiring the next generation of space explorers.

Whether you are a student of space exploration, an aspiring astronaut, or simply fascinated by human ingenuity, this book offers a thorough and accessible exploration of one of humanity's greatest accomplishments. Each chapter is meticulously crafted to highlight both the technological milestones and the human spirit that made the ISS possible.

Extras for the Curious Mind:

- **Glossary of Key Terms** – Breaks down complex space and engineering terminology for easier understanding.
- **Timeline of the ISS's Construction and Milestones** – Provides a chronological overview of the ISS's development, from conception to completion.
- **Appendices on ISS Partners, Contributions, and Technologies** – Detailed insights into the international partnerships, life-support systems, key technologies developed, and the most significant scientific experiments conducted aboard the ISS.

Join us in uncovering the legacy of the International Space Station—a living laboratory that has not only expanded our scientific horizons but also united us in the shared quest for knowledge beyond Earth. This is your invitation to explore how human collaboration in space continues to drive innovation, shape the future of space exploration, and inspire generations to come.

Dr. Leo Lexicon is an educator and author. He is the founder of Lexicon Labs, a publishing imprint that is focused on creating entertaining and educational books for active minds.

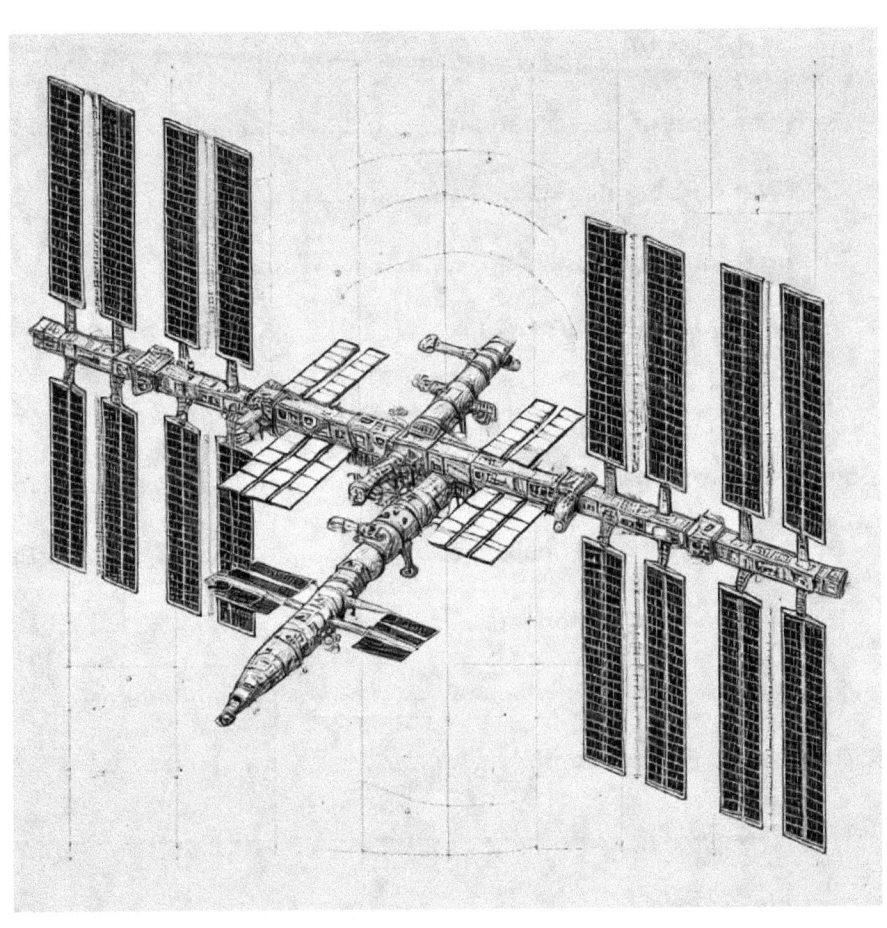

CONTENTS

Chapter 1 1

Dawn of the ISS Vision 1

    Early Concepts of Space Stations 1

    Cold War Collaboration 5

    The International Partnership 7

    The Challenges of Early Planning 9

Chapter 2 13

Engineering the Impossible 13

    Overcoming Technical Challenges 13

    Building the Key Components 16

    The Logistics of Assembly 18

    Testing and Preparations 21

Chapter 3 25

Assembly in Orbit 25

    The First Steps in Space 25

    Astronauts as Constructors 27

    Setbacks and Solutions 30

    Completing the Framework 33

Chapter 4 37
Life on the ISS 37

The Day-to-Day Life of Astronauts 37

Research and Experiments 40

Cultural and Psychological Aspects 43

Health and Safety Measures 45

Chapter 5 50
Scientific Contributions 50

Medical and Biological Research 50

Technological Innovations 53

Earth and Space Observations 55

International Research Collaborations 57

Chapter 6 61
Challenges and Crises 61

Technical Maintenance and Repairs 61

Dealing with Space Hazards 64

Political and Budgetary Struggles 66

Crisis Management on Board 69

Chapter 7 74
Legacy and Future 74

A Symbol of Peaceful Collaboration 74

The End of an Era: Future Plans 77

Preparing for Deep Space Exploration 79

Inspiring Future Generations 82

Appendices 86

A: Timeline 87

B: ISS Partners and Contributions 88

C: Life-Support Systems on the ISS 90

D: Significant ISS Experiments 91

E: Major ISS Spacewalks 91

F: ISS Educational Initiatives 92

G: Key Technologies Developed 92

Appendix H: Glossary of Key Terms 93

I. The ISS FAQ 93

Chapter 1

Dawn of the ISS Vision

Early Concepts of Space Stations

It is the early 1900s, and in a small room in Russia, Konstantin Tsiolkovsky sits hunched over a cluttered desk, surrounded by papers filled with sketches, calculations, and visions that could easily be mistaken for science fiction. The room is quiet except for the soft scratch of his pen. Tsiolkovsky, often referred to as the father of astronautics, is lost in his imagination—his thoughts leaping from the confines of Earth's gravity to the freedom of space. His hand moves swiftly across the paper, sketching the outlines of a cylindrical station floating above Earth, with solar mirrors capturing sunlight to power it. He imagines humans living and working in orbit, propelled by rockets and sustained by systems they themselves created. It is a vision decades ahead of its time, one that lays the foundation for humanity's exploration beyond our planet.

Fig. Tsiolkovsky's Original Vision Commemorated

These early ideas of space habitation, though initially the work of dreamers and theorists, were far more than whimsical notions. They were the seeds of an ambitious future, planted long before the technology to achieve them even existed. Tsiolkovsky was not alone in his dream. Years later, Wernher von Braun, the German-American rocket scientist, picked up the thread. In the 1950s, von Braun famously conceptualized an enormous rotating wheel in space—a space station that could house up to 200 people. His vision was vivid: he saw a station spinning to generate artificial gravity, a place where humans could look down upon the blue sphere of Earth and out to the vastness of the cosmos. The world von Braun imagined was one where space exploration was not limited to brief trips but was a permanent presence above Earth's atmosphere.

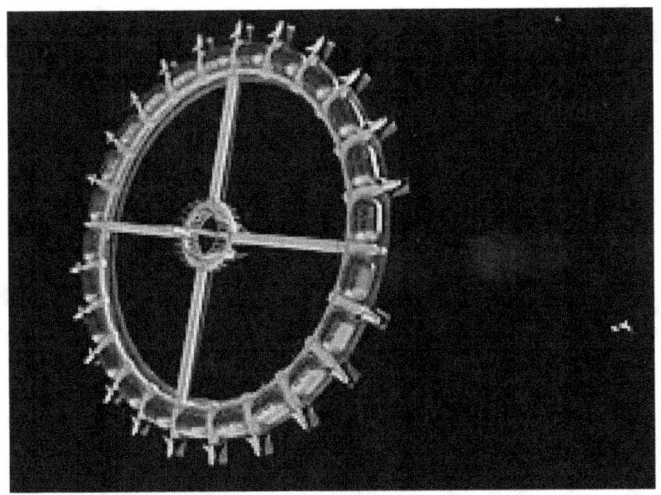

Fig. von Braun's Rotating Space Station Design
(Source: Alatorre, 2019; Creative Commons License)

Von Braun's rotating station was featured in magazines and popularized in television specials, bringing the concept of a space station into the public imagination for the first time. It was a fantastical idea, but the space race had begun, and the leap from imagination to action was closer than anyone realized. The subsequent development of Salyut and Skylab—the first actual space stations—was a direct outgrowth of these early visions. The Soviet Union launched Salyut 1 in 1971, marking the first time a manned space station was placed into orbit. A few years later, in 1973, the United States followed with Skylab, which housed astronauts for extended stays in orbit. These efforts represented humanity's first steps toward realizing the dream of sustained human presence in space, setting the stage for the grand vision that would eventually become the International Space Station.

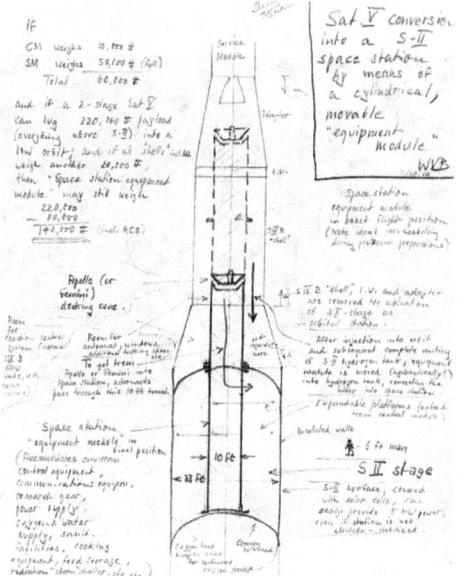

Fig. Von Braun's Design of the Space Station
(Source: Wikimedia Commons)

The seeds planted by Tsiolkovsky and von Braun were not just the stuff of imagination—they were blueprints that eventually inspired governments, scientists, and visionaries worldwide to come together. The ISS would eventually become a testament to what humanity could achieve when the brightest minds from around the world worked together. These early concepts formed the foundation of what would later be the world's most ambitious peacetime project—a project that would unite rival nations and challenge the limits of human engineering and cooperation.

Cold War Collaboration

The Cold War was a period of extreme tension, marked by an arms race and ideological rivalry between the United States and the Soviet Union. Space became a critical battleground for both nations, each desperate to prove their supremacy beyond Earth. When Yuri Gagarin became the first human in space in 1961, it was a monumental achievement for the Soviet Union. In response, the United States poured resources into its own space program, culminating in the Apollo moon landings of 1969. Space was an arena for competition—one where technological prowess equated to national pride. Yet, paradoxically, it was this very rivalry that sowed the seeds for collaboration that would eventually lead to the International Space Station.

The 1970s saw a slight thawing in the Cold War, and with it came the first real gesture of cooperation: the Apollo-Soyuz Test Project in 1975. It was the first joint U.S.-Soviet space mission, involving an American Apollo spacecraft docking with a Soviet Soyuz capsule. The mission was symbolic—a handshake in space between astronauts and cosmonauts, a gesture that proved that even the deepest of political divides could be bridged in the pursuit of knowledge. It was an early indication that the sky, and not national borders, would be the final frontier.

This moment of détente would echo into the 1980s, a time when the idea of a permanently crewed space station began to gain traction once again. In 1984, President Ronald Reagan stood before Congress and delivered a speech that would become a pivotal moment in the history of space exploration. He called for the United States to develop an international space station within a decade, describing it as an opportunity to collaborate with "freedom-loving nations." His speech captured the spirit of adventure and the prospect of unity through exploration—a stark contrast to the ideological divide of the Cold War.

While Reagan's call to action was a declaration of American leadership in space, it also opened the door to global participation. By the late 1980s and early 1990s, as the Soviet Union began to collapse, the dynamics shifted. The new Russian Federation was no longer interested in competing but rather surviving economically. This shift made collaboration not just possible but necessary. For Russia, participation in an international station provided a means to sustain its struggling space program. For the United States and its partners, bringing Russia on board was a way to capitalize on their expertise with space stations (thanks to the Mir program) and to secure broader geopolitical stability.

Thus, the Cold War—a time marked by competition and division—laid the groundwork for unprecedented collaboration. The foundation of the ISS was built upon this newfound spirit of cooperation, a stark reminder that even the most formidable rivalries could give way to partnerships when humanity looked beyond Earth. It set the stage for the international partnerships that would follow, uniting not only NASA and Roscosmos but also European, Japanese, and Canadian space agencies in a joint venture unlike any seen before.

The International Partnership

Building a space station was no small feat, and building one that involved multiple countries, each with its own interests, expertise, and limitations, not to mention the seemingly impenetrable linguistic and cultural barriers. These differences made it an even more challenging endeavor. After President Reagan's announcement, it took several years of negotiations to turn the dream of an international space station into a reality. At the heart of these discussions were five major players: NASA (United States), Roscosmos (Russia), ESA (European Space Agency), JAXA (Japan Aerospace Exploration Agency), and CSA (Canadian Space Agency).

The initial partnership agreements were a testament to diplomatic finesse. Each country had unique capabilities and expertise that they could contribute, but balancing the contributions and ensuring fair participation was a challenge. For instance, Russia, with its experience from the Mir space station, was tasked with providing essential propulsion and docking modules. The United States took the lead in overall coordination, providing power modules and the station's core structural elements. Europe contributed the Columbus laboratory, while Japan provided the Kibo laboratory, and Canada delivered its famed Canadarm2—a robotic arm crucial for assembling and maintaining the station.

The partnership was not just about dividing up construction tasks—it was also about fostering a culture of mutual respect and shared goals. Each partner had to navigate their own political landscapes back home, justifying the expenses and the risks of the venture. For instance, NASA faced scrutiny from the U.S. Congress, where some viewed the ISS as an unnecessary expense. Similarly, European partners had to contend with differing national priorities within the ESA. However, despite these challenges, the vision of an international outpost in space was compelling enough to keep the partners committed.

One of the significant breakthroughs in forging the partnership was the Intergovernmental Agreement (IGA), signed in 1998. The IGA established the legal framework that governed the station, detailing the responsibilities of each partner and the operational protocols to be followed. It was, in essence, the constitution of the ISS—a guiding document that ensured that the station would operate smoothly despite being a conglomerate of various nations and agencies. This agreement made the ISS not just a technological marvel but also a diplomatic success story, demonstrating how international cooperation could be achieved through careful negotiation and shared ambition.

The formation of the international partnership was a cornerstone in the ISS story. It reflected humanity's willingness to look beyond borders, to pool knowledge and resources for the greater good. It was not just about building a space station; it was about building trust, cooperation, and a sense of shared destiny among nations.

The Challenges of Early Planning

As ambitious as the idea of the International Space Station was, its realization was fraught with challenges. Bringing together different countries meant merging different engineering standards, design philosophies, and even languages. The initial planning phase exposed the complexity of international collaboration on such a grand scale. Engineers had to make sense of distinct systems—American and Russian units of measurement, differing power standards, and unique construction techniques. How do you ensure that a module designed in Japan would seamlessly integrate with one built in Italy, or that a Russian propulsion system would work flawlessly with an American power module?

The technical challenges were staggering. Consider the task of creating a unified life-support system. Each partner had their own approach to providing clean air, potable water, and waste recycling—systems essential for sustaining life in orbit. Integrating these various systems required innovative solutions and compromises. Communication was another major challenge. The ISS needed a unified command structure, yet astronauts from different countries spoke different languages. English and Russian were established as the official languages of the ISS, and every astronaut was required to be fluent in both—a necessity that underscored the commitment to true collaboration.

In addition to technical hurdles, the financial cost of building and maintaining the ISS was monumental. The United States bore a significant portion of the cost, but financial contributions from other partners were essential. Political support was not always guaranteed. In the United States, funding for the ISS was nearly cut several times due to budget constraints and shifting political priorities. It took lobbying from NASA, along with support from international partners, to keep the program on track. Similarly, Russia, emerging from the collapse of the Soviet Union, faced economic instability, which jeopardized its ability to contribute. Despite these challenges, the partners found ways to move forward—driven by the shared belief that the ISS was a project worth pursuing.

The early planning phase of the ISS was a balancing act—a complex dance involving technical innovation, political negotiation, and financial juggling. It required a level of international cooperation that had never been attempted before. The challenges were immense, but the determination to overcome them was even greater. Each hurdle cleared brought the world closer to achieving the dream of a permanent human presence in space, and each compromise and solution laid the foundation for the ISS as we know it today.

Let us end this Chapter with ten little-known facts about the International Space Station that highlight its unique quirks and the human side of space exploration:

1. **International Cuisine in Space**: The ISS has a menu featuring dishes from all partner countries, including Japanese sushi, Russian borscht, and American barbecue.
2. **The Most Expensive Structure Ever Built**: The ISS is estimated to have cost over $150 billion, making it the most expensive man-made structure.
3. **The Speed of Sunsets**: Astronauts on the ISS witness 16 sunrises and sunsets each day, as the station orbits Earth at about 28,000 km/h.
4. **Space Mail**: The ISS has its own postal address, and while it is mostly symbolic, letters have been delivered to the astronauts on board.
5. **Floating Musical Instruments**: There is a guitar on the ISS, and astronauts often play music to relax—Chris Hadfield's rendition of David Bowie's "Space Oddity" became an internet sensation.

6. **International Friendship**: The crew members celebrate each partner nation's holidays, sharing cultural traditions and foods.
7. **Water Recycling**: The ISS has a system that recycles urine into drinkable water—"yesterday's coffee is tomorrow's coffee," as the astronauts like to say.
8. **Space Garden**: The ISS has a small garden where astronauts grow lettuce and flowers, studying how plants grow in microgravity.
9. **Constant Maintenance**: On average, astronauts spend two hours each day maintaining the station—fixing equipment, cleaning filters, and ensuring all systems run smoothly.
10. **Weightless Sports**: The crew has fun with weightless "sports" like space ping-pong using a water droplet and small paddles.

These quirky details show that beyond the high-tech systems and rigorous science, the ISS is also a place of human connection, creativity, and adaptability—a testament to the resilience and inventiveness of the people who live and work there.

Chapter 2

Engineering the Impossible

Overcoming Technical Challenges

Creating the International Space Station was more than just a feat of human ambition; it was a testament to overcoming complex engineering challenges that pushed the boundaries of what was thought possible. The ISS needed to be modular, meaning it would be constructed in space piece by piece, with components built by different countries all needing to work seamlessly together. This ambition required technological innovation on a scale never attempted before. Engineering teams from across the world had to design systems that could be launched separately, withstand the stress of space travel, and be connected in the vacuum of space.

A significant challenge was achieving compatibility between components produced in different countries. The ISS modules were built by various international partners, each using their own engineering standards and approaches. Integrating these elements required meticulous planning. For example, the United States and Russia had different units of measurement, different design philosophies, and even different types of electrical systems. One of the first steps in overcoming these challenges was agreeing on a standardized system of specifications, such as unified docking mechanisms, electrical interfaces, and data communication protocols. This standardization was essential to ensure that modules developed in Japan, Europe, Russia, or the United States would all function as intended once assembled in space.

The challenges did not stop with standardization. The structural design of the ISS had to contend with the hostile environment of space, which includes extreme temperatures, radiation, and micrometeoroid impacts. The station would orbit Earth at a speed of about 28,000 kilometers per hour, meaning that even small debris could cause significant damage upon collision. Engineers developed specialized shielding, called Whipple shielding, to protect the station. This consisted of multiple layers that would break apart any impacting debris, dissipating the energy before it could puncture the main modules. Protecting the crew and equipment from radiation was another priority, especially during solar flare activity. The design incorporated radiation-hardened modules and dedicated "safe zones" where crew members could shelter during high radiation periods.

One of the most challenging aspects was creating a life-support system that could sustain human life for extended periods. The ISS needed to recycle air and water, provide an atmosphere similar to that of Earth, and manage human waste—all while in microgravity. The Environmental Control and Life Support System (ECLSS) was designed to recycle up to 90% of water on board, including urine, into potable water. Engineers had to create sophisticated filtration and purification systems capable of working in space conditions. Maintaining a stable temperature was also critical; external temperatures could vary from -157 degrees Celsius in the shadow of Earth to 121 degrees Celsius in direct sunlight. The ISS was equipped with radiators to dissipate excess heat and insulation to protect it from the cold.

The international nature of the project also meant that communication and command systems needed to be designed for seamless integration. Engineers had to develop a central command structure that could coordinate between the modules, all while ensuring redundancy in case of failures. This redundancy meant that vital systems, such as power and life-support, had backups that could take over without loss of functionality. The need for multiple languages also influenced design; English and Russian became the working languages of the ISS, and control systems had to be understandable by all astronauts regardless of nationality. These overlapping challenges of compatibility, durability, sustainability, and communication made the engineering of the ISS one of the most complex undertakings in human history.

Building the Key Components

The construction of the ISS relied on the development of key components that each played a vital role in creating a livable, functional space station. The modules and structural elements were designed to be launched individually and assembled in orbit, with each contributing a specific function or capability to the station. Among the first modules to be launched was Zarya, a Russian-built module whose name means "dawn." Zarya was launched in 1998 and provided essential power and propulsion capabilities in the early stages of assembly. It was the backbone that would support the first stages of the ISS, setting the foundation for subsequent modules.

Shortly after Zarya, the Unity module, built by the United States, was launched and attached to Zarya. Unity served as the connecting node for other modules and marked the beginning of true international collaboration in orbit. Unity featured six docking ports, allowing it to link with other components that would be launched in the future. The Harmony module, launched later, further expanded the station's docking capabilities and provided additional space for scientific research and crew quarters. Each module was a small but critical step toward the complete ISS, adding new capabilities and expanding the living and working area for astronauts.

One of the most notable components was the Columbus Laboratory, built by the European Space Agency. Columbus was Europe's primary scientific contribution, launched in 2008. It featured a fully equipped laboratory where experiments in biology, physics, and materials science could be conducted in microgravity. The Japanese Kibo module, launched the same year, represented Japan's significant contribution. Kibo was the largest module on the ISS and included an external platform where experiments could be exposed to the vacuum of space. This unique feature allowed for studies on the effects of the space environment on various materials and biological samples.

The Canadarm2, contributed by the Canadian Space Agency, was another key component that was essential for assembling the ISS. It was an advanced robotic arm capable of moving large modules and aiding astronauts during spacewalks. Canadarm2 provided the dexterity needed to attach modules, install equipment, and even capture uncrewed cargo vehicles approaching the station. The construction of the ISS also included solar arrays, which provided the power necessary to run all of the station's systems. The large, wing-like solar panels were designed to rotate to follow the Sun, maximizing energy collection as the station orbited Earth.

These key components were built over many years and in different countries, but they were all designed with the ultimate goal of being able to function as a cohesive unit once in space. The complexities of designing, testing, launching, and assembling these components required unprecedented coordination among the international partners. Each partner contributed according to their expertise and technological capabilities, and each component brought the ISS closer to being a fully operational space laboratory. The building of these modules was not just an exercise in engineering but also a logistical challenge, as each piece had to be launched, connected, and activated in the precise order necessary to ensure the growing station's stability and functionality.

The Logistics of Assembly

Assembling the ISS was akin to putting together a massive puzzle, with each piece being launched separately, often years apart, and fitted together in the microgravity of space. The logistics of assembly involved meticulous planning and the coordinated efforts of numerous space agencies. Every module and component had a precise place and role, and they all had to be launched into space by rockets or carried by the Space Shuttle, which acted as the workhorse of ISS construction in its early years.

The role of the Space Shuttle was fundamental in the assembly of the ISS. The shuttle missions transported not only the various modules but also the astronauts who would conduct spacewalks, known as extravehicular activities (EVAs), to connect the modules and perform essential construction tasks. These missions required detailed choreography, with shuttle crews spending months in training, rehearsing the exact procedures that would be carried out in space. Spacewalks, sometimes lasting over eight hours, were the cornerstone of the construction process. Astronauts had to bolt modules together, connect power lines, set up data links, and perform countless other tasks—all while working in bulky spacesuits, navigating the challenges of zero gravity, and contending with the harsh environment of space.

The Russian Proton rocket was another crucial vehicle for transporting modules to space. The Proton rocket launched several key modules, such as Zarya and Zvezda, which were essential for providing propulsion, navigation, and initial power to the station. Each launch required perfect timing and coordination to ensure that the modules reached the correct orbit and could be successfully attached to the existing structure. The margin for error was minimal; any mistake could result in costly delays or, worse, the loss of a module. The international nature of the project added another layer of logistical complexity, with each launch requiring collaboration between different mission control centers, including NASA in

Houston, Roscosmos in Moscow, and ESA's control center in Germany.

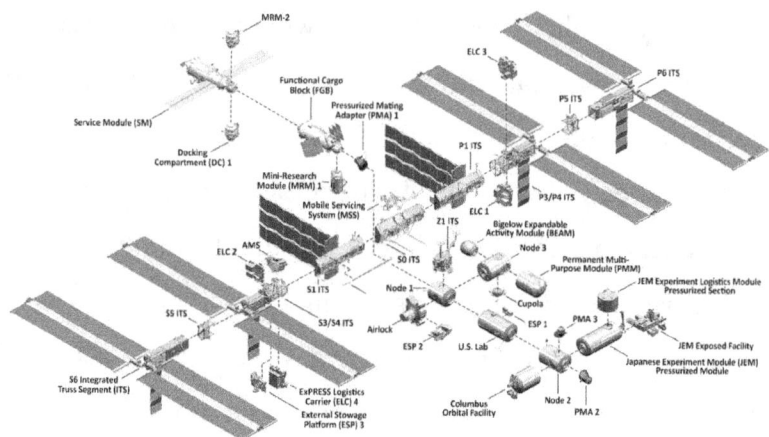

Fig. Exploded View of the ISS (Source: NASA)

One of the major challenges during assembly was the alignment of modules in orbit. Each module had to be precisely maneuvered into position, often with the aid of the Canadarm2, which was controlled by astronauts inside the station. The robotic arm, capable of handling heavy loads in space, was essential for docking modules that could not be manually positioned by spacewalking astronauts. The integration of these components required not only precision but also adaptability, as unexpected issues would often arise, necessitating real-time problem-solving by both the crew and mission control teams on the ground.

Cargo resupply was another critical aspect of the assembly logistics. The station needed regular deliveries of equipment, spare parts, and supplies for the crew. In the early years, the Space Shuttle served this purpose, carrying large payloads that included replacement parts for the station's growing array of systems. Later, uncrewed cargo vehicles, such as Russia's Progress spacecraft, ESA's Automated Transfer Vehicle (ATV), and Japan's H-II Transfer Vehicle (HTV), played a significant role in keeping the station stocked with essential supplies. These cargo missions ensured that the assembly process could continue smoothly and that the crew had the tools and materials they needed to keep the project on track.

The logistics of assembling the ISS were further complicated by the station's orbit, which meant that construction activities could only be conducted during certain windows when the ISS was in sunlight and in a position that allowed safe EVA operations. Despite these challenges, the station grew, module by module, from a small base consisting of just Zarya and Unity to the sprawling structure that orbits Earth today. Each successful launch and assembly milestone represented not only progress for the ISS but also a testament to the collaboration and ingenuity of thousands of engineers, technicians, and astronauts across the globe.

Testing and Preparations

Before any module of the ISS could be launched into space, it underwent rigorous testing to ensure that it could withstand the harsh conditions of space and function as intended once in orbit. The testing process was crucial, as once a module was launched, it would be nearly impossible to correct any major flaws. This phase of preparation included both ground-based testing and the training of astronauts who would assemble and live on the station.

Ground testing involved putting each module through simulations that replicated the extreme environment of space. Thermal testing was conducted to see how modules would respond to the drastic temperature fluctuations experienced in orbit. Vibration tests ensured that the modules could survive the intense shaking and forces encountered during launch. Engineers also conducted vacuum tests to ensure that the modules would maintain their integrity in the absence of atmospheric pressure. Each module's systems, such as electrical, thermal, and data communications, were tested to verify their reliability. The integrated nature of the ISS meant that every part had to be compatible with others, requiring joint tests that involved multiple international partners.

In addition to testing the hardware, astronaut training was an integral part of the preparations for ISS assembly. Astronauts trained for months, and sometimes years, to be ready for the demanding tasks required during assembly. NASA's Neutral Buoyancy Laboratory, a massive swimming pool that mimics the microgravity environment of space, was used extensively for training spacewalks. Astronauts would practice wearing their spacesuits underwater, performing the exact tasks they would carry out during EVAs. This training helped them become familiar with the tools and procedures they would use, as well as the limitations imposed by the spacesuits. The European Space Agency and the Russian Space Agency had similar facilities, and training was often a collaborative effort, with astronauts from different countries working together to prepare for their missions.

Training also involved simulations of onboard emergencies. The crew needed to be prepared for scenarios such as depressurization, fire, or system failures. These simulations, conducted in high-fidelity mock-ups of the ISS modules, helped astronauts become familiar with emergency protocols and learn how to work as a cohesive team under pressure. Language training was another crucial aspect, as all astronauts had to be proficient in both English and Russian to ensure effective communication between crew members and mission control centers. The time spent preparing for life on the ISS was intensive, aimed at ensuring that the crew would be ready to face any challenge that might arise.

The preparations extended beyond just the astronauts and hardware. The mission control teams also went through extensive rehearsals to ensure they were ready for the complexities of managing the assembly of the ISS. Mission simulations were conducted, often involving the crew and ground teams working together to practice the timing of commands, the coordination of EVAs, and the handling of unexpected issues. These simulations allowed teams to refine their procedures and ensure that they were ready for the real mission. Each phase of preparation was essential to minimize the risks involved in assembling such a complex structure in space.

Fig. The ISS in Orbit

The culmination of these testing and preparation efforts was evident in the successful launch and assembly of the ISS modules. Each module that was launched represented years of work by thousands of individuals—engineers who designed and tested the components, astronauts who trained to install them, and mission control teams who coordinated every step. The preparation process ensured that the ISS was not only functional but also a safe and livable environment for the astronauts who would call it home. The thoroughness of the testing and training programs was key to overcoming the inherent dangers of space and creating a stable platform for scientific research and international cooperation.

Chapter 3

Assembly in Orbit

The First Steps in Space

The launch of the first components of the International Space Station marked the beginning of one of humanity's greatest achievements in space. It all started in November 1998 with the launch of Zarya, the first module of the ISS. The name Zarya, meaning "dawn" in Russian, was fitting—it represented the dawn of a new era of international collaboration in space exploration. The launch of Zarya by a Russian Proton rocket from Baikonur Cosmodrome was the culmination of years of planning and engineering. It was a symbol of international collaboration in a post-Cold War world, where former adversaries came together to pursue the shared dream of a permanent human outpost in orbit.

Zarya was a multifunctional module, providing power, propulsion, and storage for the station's early assembly stages. It was followed in December 1998 by the launch of the Unity module aboard the Space Shuttle Endeavour. Unity, built by NASA, was a connecting node with six docking ports that would serve as the primary hub for the attachment of future modules. The shuttle mission STS-88 brought together Zarya and Unity, marking the first physical connection between Russian and American-built segments in orbit. Astronauts from the United States worked side-by-side with Russian engineers on the ground to ensure the successful docking of these two foundational modules.

The initial phase of assembly was marked by intense collaboration and meticulous coordination. Every step of the process was a high-stakes operation; each launch, each docking, and each connection was an opportunity for success or a risk of failure. The complexities of working in microgravity required innovations in tools and procedures. Spacewalking astronauts, in bulky suits, had to carry out tasks that demanded both precision and strength. These early missions, which connected Zarya and Unity, were essential in creating a foundation upon which the rest of the ISS could be built. The assembly process was not without its challenges. In the early stages, the station consisted of only a few modules, and power and propulsion capabilities were limited. The coordination between the modules required careful control of power sharing and alignment to maximize efficiency. Crew members aboard the Space Shuttle and ground control teams worked tirelessly to ensure that the systems were functioning properly and that each successive step of construction was completed without setbacks. The early success of connecting Zarya and Unity demonstrated that international collaboration in orbit was not only possible but could be achieved smoothly despite the complexities involved.

In July 2000, the Zvezda Service Module was launched by a Russian Proton rocket, adding essential living quarters, life-support systems, and propulsion capabilities. Zvezda was the heart of the station, designed to provide the amenities necessary for a permanent human presence in space, including sleeping quarters, a galley, and an oxygen generator. Its addition marked the point at which the ISS became habitable, allowing astronauts to begin long-duration stays aboard the station. This was a significant milestone, as it transitioned the ISS from a collection of connected modules to a livable environment for astronauts from around the world.

Astronauts as Construction Experts

The construction of the ISS was unlike any construction project ever undertaken by humanity. It required astronauts to serve as both explorers and builders, working in the most challenging environment imaginable—space. The process of assembly was an ongoing endeavor, with each crew adding new components, connecting electrical and data lines, and ensuring that all systems functioned as one cohesive unit. The astronauts, working outside the station during extravehicular activities (EVAs), played a crucial role in making the ISS what it is today.

Spacewalks were the cornerstone of ISS assembly. Astronauts had to perform a wide variety of tasks while floating in the vacuum of space, tethered to the station and relying on their training and teamwork to ensure success. Each spacewalk was a meticulously planned operation, often rehearsed in the Neutral Buoyancy Laboratory—a massive pool used to simulate weightlessness on Earth. These rehearsals helped astronauts become familiar with the specific tools and procedures they would use during their EVAs. Despite this preparation, working in space was far from easy. The bulky suits made movement cumbersome, the vacuum of space presented constant danger, and the tasks at hand often required fine motor skills, such as bolting modules together or connecting delicate cables.

The first major assembly spacewalks took place during the STS-88 mission, when astronauts Jerry Ross and Jim Newman connected cables between Zarya and Unity. This work was vital for integrating power and data systems between the modules. The spacewalks during these early missions were lengthy and physically demanding.

Astronauts had to move hand-over-hand along the station's exterior, carefully navigating the module surfaces while securing themselves with tethers to prevent drifting away. They carried specialized tools designed for use in microgravity, where even simple tasks like turning a wrench require effort to avoid spinning oneself instead of the tool.

The use of the Canadarm2, a robotic arm operated from inside the station, was another critical aspect of construction. The Canadarm2, provided by the Canadian Space Agency, allowed astronauts to maneuver large modules and equipment into place with precision that would be impossible to achieve manually. Spacewalkers and the robotic arm operators had to work in perfect synchrony to install components like solar arrays, radiator panels, and additional laboratories. This coordination required constant communication and teamwork, with astronauts relying on each other and on ground control to guide the arm, verify alignments, and secure each piece.

The addition of new modules required a consistent process of verification and integration. Astronauts often had to conduct multiple spacewalks to complete the installation of a single component, ensuring that all power, thermal, and data connections were secure. One particularly challenging task was the installation of the solar arrays, which were crucial for powering the growing station. These massive arrays had to be carefully unfolded and connected, with astronauts troubleshooting issues like misaligned panels or jammed mechanisms. The work was painstaking, but it was through these efforts that the ISS became self-sufficient, able to generate enough power to support both the crew and the myriad of experiments onboard.

The process of assembling the ISS was a testament to human ingenuity and determination. The astronauts serving as constructors were not just following instructions; they were problem solvers, often dealing with unexpected challenges in real time. The environment of space, with its extreme temperatures, zero gravity, and the ever-present threat of micrometeoroids, was unforgiving. Yet, through their dedication and skill, these astronauts turned individual modules into a functioning space station that has supported human life continuously since the early 2000s.

Setbacks and Solutions

The assembly of the International Space Station was far from a flawless process. Numerous setbacks tested the resilience and ingenuity of the teams involved. From technical failures to environmental hazards, each challenge required creative problem-solving and international cooperation to overcome. One of the earliest challenges arose shortly after the launch of the Zvezda module. Zvezda's integration was critical for providing life-support systems, but shortly after docking, the crew discovered alignment issues that affected power connections between the modules.

Engineers on the ground had to work quickly to develop new procedures, and astronauts performed spacewalks to manually adjust and secure the connections. It was a tense period, but through teamwork and adaptability, the modules were successfully integrated.

Another major setback occurred during the deployment of the solar arrays. These arrays were essential for powering the expanding station, but during an early attempt to extend one of the arrays, it became partially jammed. The deployment mechanism had malfunctioned, causing part of the array to snag and preventing it from fully unfurling. The crew aboard the station had to work with mission control to carefully analyze the situation and determine a solution. Ultimately, it required an unplanned spacewalk, during which astronauts manually freed the array. The operation was delicate, as any mistake could have caused irreversible damage to the panels or risked the safety of the astronauts involved. The success of the mission was a testament to the astronauts' training and the support from mission control teams on Earth.

Space debris presented another constant challenge. The ISS, orbiting at over 28,000 kilometers per hour, faced potential impacts from micrometeoroids and fragments of old satellites. The station was designed with shielding to protect against small debris, but larger objects still posed a significant risk. In several instances, the crew had to take shelter in the Soyuz escape module when debris was predicted to come dangerously close to the station. These "debris conjunction" warnings required quick decision-making and coordination between NASA, Roscosmos, and other international partners to track the debris and determine the safest course of action. Mechanical failures also complicated the construction and operation of the ISS. One notable incident involved the cooling system, which is crucial for maintaining a stable temperature for both equipment and crew. In 2010, a critical ammonia pump module failed, severely impacting the station's ability to regulate its temperature. The crew had to perform a series of complex spacewalks to replace the failed pump. The repair was difficult due to the size and weight of the pump, as well as the intricacies of connecting ammonia lines in the vacuum of space. The ammonia itself was hazardous, and any leak could have posed a significant danger to the astronauts. Despite these risks, the repair was successfully completed, restoring the cooling system and ensuring the continued operation of the station.

The setbacks faced during the ISS assembly were not limited to technical issues; political and financial hurdles also played a significant role. The international nature of the project meant that changes in government priorities or economic conditions in any of the partner countries could threaten progress. For example, budget constraints in Russia during the 1990s led to delays in the production and launch of key modules. The international partners had to work together to adjust timelines, redistribute resources, and ensure that the project stayed on track. This collaboration was a critical aspect of overcoming the challenges, demonstrating that the ISS was as much a political achievement as it was an engineering one.

Each setback encountered during the construction of the ISS required a solution that involved not just the astronauts in space, but also the engineers, scientists, and mission control teams on the ground. It was a true demonstration of international teamwork and resilience. The ability to adapt to unexpected situations, develop new procedures on the fly, and maintain the safety of the crew was what ultimately allowed the ISS to grow from a few small modules into the expansive, fully operational space station that it is today.

Completing the Framework

With the early modules in place and numerous setbacks overcome, the focus shifted to completing the framework of the International Space Station. The assembly of the ISS was a modular process, requiring the sequential addition of laboratories, living quarters, and external trusses that provided support for solar arrays and radiators. Completing this framework was a major engineering milestone, transforming the ISS from a collection of interconnected modules into a fully functional space laboratory capable of supporting long-term human habitation and complex scientific research.

One of the major milestones in completing the framework was the installation of the Integrated Truss Structure (ITS). The truss system was critical for the stability and functionality of the ISS, serving as the backbone to which solar arrays, radiators, and other external equipment were attached. The ITS was delivered and installed over multiple shuttle missions, with each segment adding new capabilities to the station. The truss system allowed for the expansion of power generation, with additional solar arrays increasing the station's capacity to support more experiments and larger crews. These solar arrays, each spanning over 70 meters when fully extended, were installed and deployed by astronauts during spacewalks. The precise positioning of the truss segments was crucial, as even a slight misalignment could affect the station's balance and power generation efficiency.

The addition of laboratories from different international partners marked another significant phase in completing the ISS framework. The Destiny Laboratory, launched in 2001, was NASA's primary research module and the first major U.S. laboratory on the ISS. Destiny provided a state-of-the-art facility for conducting a wide range of experiments, from studying the effects of microgravity on the human body to testing new materials and technologies. The European Space Agency's Columbus Laboratory and Japan's Kibo Laboratory further expanded the station's scientific capabilities. Each of these laboratories was designed to accommodate experiments that could not be conducted on Earth, taking advantage of the unique environment provided by the ISS.

The Node modules, including Harmony and Tranquility, played an essential role in completing the framework of the ISS. These nodes served as connecting hubs, allowing additional modules to be attached and providing critical infrastructure for life-support, power distribution, and crew quarters. Harmony, launched in 2007, provided docking ports for European and Japanese laboratories, while Tranquility, launched in 2010, housed life-support systems and crew living quarters. Tranquility also included the Cupola, a multi-windowed observation module that gave astronauts an unparalleled view of Earth and the surrounding space. The Cupola became a favorite spot for astronauts, offering a place to observe and photograph the planet, enhancing both the scientific and emotional experience of living in space.

The logistics of completing the ISS framework required careful coordination of resources and schedules. Each module had to be launched in a specific order, with new additions dependent on the successful installation of previous components. The Space Shuttle played a critical role in this process, delivering modules, truss segments, and equipment needed for assembly. The shuttle missions were planned with precision, often carrying multiple payloads that needed to be installed during the same mission. Astronauts aboard the shuttle and the ISS worked together to ensure that each component was successfully attached, integrated, and activated.

One of the final major components to be added was the Alpha Magnetic Spectrometer (AMS-02), a cutting-edge particle physics experiment designed to study cosmic rays and search for evidence of dark matter and antimatter. The AMS was installed on the ISS in 2011, marking the completion of the major scientific facilities on the station. The installation of the AMS represented the culmination of over a decade of work to transform the ISS into a world-class research facility. The AMS, along with the other laboratories and experimental platforms, ensured that the ISS would continue to make significant contributions to science and technology for years to come.

Completing the framework of the ISS was not just about adding new modules; it was about creating a sustainable, adaptable platform that could support a wide range of activities, from scientific research to international diplomacy. Each new addition brought new capabilities, whether it was increased power generation, expanded living space, or enhanced research opportunities. The ISS became a symbol of what could be achieved when nations worked together toward a common goal. The completed station was a marvel of engineering, a testament to human ingenuity, and a platform for pushing the boundaries of what is possible in space exploration and scientific discovery.

Chapter 4

Life on the ISS

The Day-to-Day Life of Astronauts

Life on the International Space Station is a unique blend of science, engineering, and human endurance. Astronauts aboard the ISS must adapt to a daily routine that is unlike anything experienced on Earth. The day-to-day activities are carefully structured to ensure that crew members maintain their physical and mental well-being while conducting important scientific research. Each day begins with a wake-up call from Mission Control, followed by a briefing on the day's schedule. The environment on the ISS requires a high level of discipline, as every aspect of life is influenced by the station's microgravity conditions.

THE INTERNATIONAL SPACE STATION

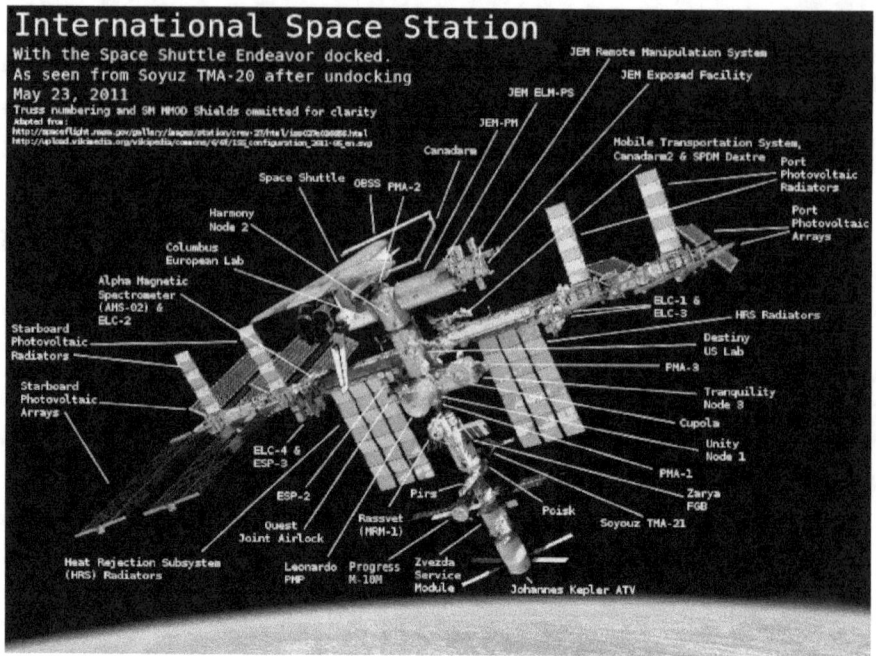

Fig. ISS with the Space Shuttle Docked (Source: NASA)

One of the major differences between life on Earth and life aboard the ISS is the absence of gravity. Microgravity affects everything, from eating to sleeping to exercising. Astronauts sleep in small compartments, where they strap themselves into sleeping bags to prevent floating around while asleep. They have to adjust to the sensation of floating throughout the station, which means that simple activities, such as moving from one module to another, require using handrails and foot straps. Adapting to microgravity takes time, and astronauts often experience "space adaptation syndrome," a form of motion sickness that occurs during the first few days in orbit.

Daily routines on the ISS are designed to strike a balance between work, exercise, and leisure. Crew members spend several hours each day conducting scientific experiments, maintaining the station, and carrying out technical tasks such as repairs. Scientific research is a primary focus, with experiments spanning a wide range of disciplines, including biology, physics, and materials science. Astronauts may study the effects of microgravity on the human body, grow plants to test agricultural techniques, or observe physical phenomena that cannot be replicated on Earth. The results of these experiments have implications not only for space exploration but also for advancements in medicine, agriculture, and technology back home.

Physical exercise is another critical component of the daily routine. In microgravity, muscles and bones do not experience the same forces they do on Earth, leading to muscle atrophy and bone density loss. To combat these effects, astronauts are required to exercise for approximately two hours each day. The station is equipped with exercise machines such as the Treadmill with Vibration Isolation and Stabilization (TVIS), a stationary bicycle, and the Advanced Resistive Exercise Device (ARED), which allows astronauts to perform weightlifting exercises in a weightless environment. Exercise helps maintain cardiovascular health, muscle strength, and bone density, enabling astronauts to return to Earth without severe physical impairments.

Leisure time is limited, but essential for the mental well-being of the crew. Astronauts have access to a variety of activities to help them relax, including watching movies, reading books, and communicating with family members via video calls or email. Observing Earth from the Cupola—a multi-windowed module with breathtaking views of the planet below—is a favorite pastime for many astronauts. This unique perspective allows them to see Earth from a completely different vantage point, watching sunrises and sunsets, observing weather patterns, and appreciating the beauty of the planet. These moments of reflection help crew members stay connected to home, despite being hundreds of kilometers above the Earth's surface.

Research and Experiments

The International Space Station serves as one of the most unique laboratories ever created. The microgravity environment allows scientists to conduct experiments that are impossible to replicate on Earth, providing valuable insights into a wide range of fields, from human health to materials science. The research conducted aboard the ISS is designed to advance knowledge not only for space exploration but also for applications that benefit humanity as a whole.

One key area of research on the ISS is the study of the effects of microgravity on the human body. Prolonged exposure to microgravity presents a number of health challenges, including muscle atrophy, bone density loss, and changes in the cardiovascular system. The ISS provides a platform for studying these effects in detail, helping scientists develop countermeasures that can be used to protect astronauts during long-duration space missions, such as a potential journey to Mars. Experiments have also provided insights into the immune system, which behaves differently in space, leading to changes in how the body responds to infections. Understanding these changes is crucial for ensuring the health of astronauts on future missions and can also lead to breakthroughs in medical treatments on Earth.

Beyond human health, the ISS is home to a variety of biological experiments, including the study of plants in space. Growing plants in microgravity is challenging, as the absence of gravity affects how water and nutrients move through the soil and how roots develop. Researchers aboard the ISS have successfully grown a number of plants, including lettuce, radishes, and even flowers, helping to determine which techniques work best for cultivating crops in space. This research is essential for future space missions, where astronauts may need to grow their own food during extended journeys. The results of these experiments could also have applications for agriculture on Earth, particularly in environments where traditional farming methods are not feasible.

The physical sciences are another area of focus for ISS research. The microgravity environment provides an opportunity to study phenomena that are obscured by gravity on Earth. For instance, fluid dynamics are fundamentally different in space, as fluids do not settle in the same way without gravity pulling them down. Researchers have conducted experiments on fluid behavior, which have applications in industries such as oil extraction and manufacturing. Combustion experiments aboard the ISS have also provided valuable insights, as flames burn differently in microgravity. Understanding how fire behaves in space can help improve fire safety for future missions and can also lead to more efficient combustion processes for use on Earth.

The ISS also serves as a platform for technological innovation. The challenges of living and working in space have driven the development of new technologies that have practical applications on Earth. One such example is the advancement of water recycling systems. The Environmental Control and Life Support System (ECLSS) on the ISS recycles water, including urine, into potable water, using a combination of filtration, distillation, and chemical treatments. These water recycling technologies are being adapted for use in areas of the world where access to clean drinking water is limited. The research and experiments conducted aboard the ISS demonstrate the power of space exploration to drive innovation and improve life on Earth.

Cultural and Psychological Aspects

Living aboard the International Space Station requires astronauts to adapt not only physically but also psychologically. The unique environment of the ISS presents a number of psychological challenges, including isolation, confinement, and separation from family and friends. The cultural diversity of the crew adds another layer of complexity, as astronauts from different countries must work together as a cohesive team despite differences in language, culture, and customs. The ISS is a true microcosm of international cooperation, where astronauts from around the world come together to live and work in harmony.

The psychological challenges of living in space are significant. Astronauts are isolated from the rest of humanity, confined to a relatively small space for months at a time, and face the constant pressure of living in a hazardous environment. The lack of privacy can also be challenging, as crew members share living quarters, workspaces, and even sleeping areas. Maintaining mental health is a priority for space agencies, and astronauts undergo extensive psychological training before their missions to prepare for the challenges they will face. Regular communication with family and friends is encouraged, and video calls, emails, and messages help astronauts stay connected to their loved ones on Earth.

Team dynamics are an essential aspect of life aboard the ISS. The crew is made up of astronauts from different countries, each bringing their own cultural background, language, and work style. Effective communication is crucial, and all crew members are required to be proficient in both English and Russian, the two official languages of the ISS. This linguistic requirement helps ensure that everyone can communicate effectively, both with each other and with mission control centers in the United States and Russia. Cultural training is also an important part of astronaut preparation, helping crew members understand and appreciate the customs and traditions of their international colleagues.

Despite the challenges, the cultural diversity of the ISS crew is one of its greatest strengths. Astronauts bring unique perspectives and experiences to the mission, and the opportunity to work with people from different backgrounds fosters a spirit of cooperation and mutual respect. The crew celebrates cultural events and holidays from each of their home countries, sharing traditional foods and customs. These celebrations provide a sense of normalcy and help boost morale, creating a positive atmosphere aboard the station. The ability to work together effectively, despite differences, is a testament to the power of international collaboration and a model for how people from different cultures can come together to achieve a common goal.

Astronauts also rely on routines and hobbies to help maintain their mental well-being. Exercise, as mentioned earlier, plays a crucial role in maintaining physical health, but it also helps alleviate stress and improve mood. Crew members are encouraged to pursue hobbies during their leisure time, whether it is reading, drawing, playing musical instruments, or simply watching movies. The Cupola, with its panoramic views of Earth, offers a place for reflection and relaxation. Many astronauts describe the experience of looking down at Earth as deeply moving, providing a sense of perspective and reminding them of the interconnectedness of all life on the planet.

Health and Safety Measures

The unique environment of the International Space Station requires extensive health and safety measures to protect the crew from the various hazards of space. Maintaining the health of astronauts is a top priority for space agencies, as the risks of illness or injury are amplified in the microgravity environment. The ISS is equipped with a range of medical supplies, and crew members undergo rigorous medical training to handle potential emergencies. The station's systems are designed with multiple redundancies to ensure the safety of the crew, and detailed protocols are in place for dealing with emergencies such as fire, depressurization, or medical issues.

The microgravity environment presents a number of health challenges for astronauts. One of the most significant effects of prolonged exposure to microgravity is the loss of bone density and muscle mass. Without the constant pull of gravity, bones lose minerals, and muscles atrophy, which can lead to weakness and an increased risk of fractures. To mitigate these effects, astronauts follow a strict exercise regimen that includes both cardiovascular and resistance training. The Advanced Resistive Exercise Device (ARED) allows astronauts to simulate weightlifting exercises, helping to maintain muscle mass and bone density. Exercise is not only important for physical health but also plays a role in maintaining mental well-being, providing a sense of routine and accomplishment.

Radiation is another major health concern for astronauts aboard the ISS. The station orbits Earth at an altitude where the protective effects of the planet's atmosphere are significantly reduced, exposing the crew to higher levels of cosmic radiation. Solar flares and other space weather events can lead to spikes in radiation levels, posing a risk to the health of the crew. The ISS is designed with shielding to protect against radiation, and astronauts have designated "safe zones" within the station where they can take shelter during periods of increased solar activity. Dosimeters are used to monitor radiation exposure, and mission duration is limited to reduce the cumulative radiation dose received by each astronaut.

Medical emergencies are a significant concern in space, as the crew is far from the nearest hospital. The ISS is equipped with a Health Maintenance System (HMS), which includes medical supplies and equipment for dealing with a range of medical issues, from minor injuries to more serious conditions. Crew members receive extensive medical training before their missions, learning how to use the medical equipment onboard and how to perform basic procedures. Telemedicine also plays a crucial role in health care on the ISS, allowing astronauts to consult with doctors on Earth in real time. In the event of a serious medical emergency, the crew has the option to return to Earth aboard the Soyuz spacecraft, which serves as an emergency escape vehicle.

Fire is one of the most dangerous hazards aboard the ISS. The confined environment of the station, combined with the presence of flammable materials and electrical systems, means that a fire could spread rapidly if not contained. The ISS is equipped with smoke detectors, fire extinguishers, and emergency masks, and the crew conducts regular fire drills to ensure they are prepared to respond quickly in the event of a fire. The unique behavior of flames in microgravity adds an additional layer of complexity to fire safety, as flames tend to form spherical shapes and can burn at lower temperatures. Understanding how fire behaves in space is crucial for developing effective fire suppression techniques and ensuring the safety of the crew.

Depressurization is another critical emergency scenario that the crew must be prepared for. A breach in the station's hull, whether caused by a micrometeoroid impact or a mechanical failure, could lead to a rapid loss of air pressure. The ISS is designed with multiple layers of shielding to protect against impacts, and the station's modules can be sealed off from one another in the event of a breach. The crew conducts regular drills to practice locating and sealing leaks, and all crew members are trained in the use of emergency oxygen masks and pressure suits. The ability to respond quickly and effectively to a depressurization event is essential for ensuring the safety of the crew.

The health and safety measures aboard the ISS are a testament to the careful planning and engineering that have gone into creating a habitable environment in one of the most challenging places imaginable. The unique hazards of space require innovative solutions and constant vigilance, and the protocols in place help ensure that the crew can live and work safely while conducting the important research that makes the ISS a valuable asset for humanity. The combination of medical training, safety drills, advanced technology, and international cooperation has allowed astronauts to thrive in an environment that, just a few decades ago, was considered beyond the reach of sustained human habitation.

DR. LEO LEXICON

Chapter 5

Scientific Contributions

Medical and Biological Research

The International Space Station has served as an unparalleled platform for medical and biological research, advancing our understanding of human physiology, cellular biology, and disease mechanisms. The microgravity environment presents unique conditions that allow scientists to investigate biological processes in ways not possible on Earth. The findings from these experiments have implications for both long-duration space missions and for medical advancements that benefit people on Earth.

One of the key areas of medical research on the ISS is the study of muscle atrophy and bone density loss. Prolonged exposure to microgravity leads to significant muscle and bone deterioration, which poses a major challenge for astronauts on long-term missions. Researchers have used the ISS to study the mechanisms behind these changes, examining how the lack of gravitational force affects muscle fibers and bone cells. These studies have led to the development of countermeasures, such as specific exercise regimens and nutritional interventions, to help mitigate the effects of microgravity on the human body. Insights gained from this research have also contributed to our understanding of osteoporosis and muscle-wasting diseases on Earth, paving the way for new treatments.

In addition to musculoskeletal research, the ISS has been instrumental in studying the cardiovascular system. Microgravity causes fluid shifts within the body, which affects blood circulation, heart function, and vascular health. The Fluid Shifts experiment, for example, has helped scientists understand how these changes impact intracranial pressure and vision, a condition known as Spaceflight-Associated Neuro-ocular Syndrome (SANS). This research is crucial for ensuring astronaut health during extended missions, and it also provides valuable insights into conditions like glaucoma and hypertension, which affect millions of people on Earth.

Cellular biology is another area where the ISS has made significant contributions. Microgravity affects cell behavior, gene expression, and the immune response. Experiments conducted aboard the station have shown that certain cells, such as immune cells, behave differently in microgravity, which can lead to reduced immune function. This finding has implications for both astronaut health and for patients on Earth with compromised immune systems. Researchers have also studied how cancer cells grow in space, discovering that some tumors grow more aggressively in microgravity. These insights are helping scientists develop better models for cancer research and test new anti-cancer drugs.

Plant biology is yet another critical area of research on the ISS. Understanding how plants grow in space is essential for future missions where astronauts may need to produce their own food. Experiments with various crops, such as wheat, lettuce, and radishes, have provided insights into how microgravity affects plant growth, root development, and nutrient uptake. The Advanced Plant Habitat and Veggie experiments have helped determine optimal conditions for growing food in space, including lighting, watering, and nutrient delivery. These findings are not only crucial for space exploration but also have applications for agriculture on Earth, particularly in areas with challenging growing conditions.

Technological Innovations

The challenges of building, maintaining, and living aboard the International Space Station have driven a wide array of technological innovations. Many of these technologies, developed to overcome the unique difficulties of space, have found applications on Earth, improving industries and enhancing daily life. The ISS serves as a testing ground for technologies that not only support space exploration but also have far-reaching benefits for humanity.

One of the most impactful technological innovations developed for the ISS is the Environmental Control and Life Support System (ECLSS). This system is designed to recycle water and air, ensuring a sustainable environment for astronauts. The ECLSS can reclaim up to 90% of water used on the station, including water from sweat, urine, and other sources. The technology behind this water recycling system has been adapted for use in areas of the world where access to clean drinking water is limited. By providing a reliable source of potable water, these systems are helping to address the global water crisis and improve living conditions in underserved communities.

Power generation and energy efficiency are other areas where the ISS has driven technological advancements. The station's massive solar arrays provide all of its power, converting sunlight into electricity through photovoltaic cells. These arrays are designed to track the Sun, maximizing energy collection as the station orbits Earth. The technology used in the ISS solar arrays has contributed to advancements in solar power on Earth, making solar panels more efficient and affordable. The need for efficient power management aboard the ISS has also led to innovations in battery technology, with advanced lithium-ion batteries developed for the station now being used in electric vehicles and renewable energy storage.

Robotics has played a crucial role in the construction and maintenance of the ISS, leading to significant advancements in robotic technology. The Canadarm2, a robotic arm developed by the Canadian Space Agency, has been used to assemble modules, capture cargo vehicles, and assist astronauts during spacewalks. The technology behind the Canadarm2 has been adapted for use in various terrestrial applications, including robotic surgery. The precision and dexterity required for robotic operations in space have translated into improvements in medical robotics, allowing surgeons to perform minimally invasive procedures with greater accuracy.

The ISS has also been a platform for testing new materials and manufacturing processes. The microgravity environment allows scientists to create materials with unique properties that cannot be produced on Earth. One example is the development of fiber-optic cables with improved clarity and reduced signal loss. In microgravity, the fibers can be drawn without the imperfections that typically occur under the influence of gravity, resulting in higher-quality cables. These advancements have the potential to improve telecommunications infrastructure on Earth, enabling faster data transmission and more reliable internet connectivity.

Earth and Space Observations

The International Space Station provides a unique vantage point for observing both Earth and the cosmos. Orbiting at an altitude of approximately 400 kilometers, the ISS offers an unparalleled perspective on our planet, allowing scientists to monitor environmental changes, track natural disasters, and study atmospheric phenomena. The station is also equipped with instruments that contribute to our understanding of space, conducting astronomical observations and studying cosmic radiation.

One of the key contributions of the ISS to Earth observation is its role in monitoring climate change. The station is equipped with a variety of sensors and cameras that capture high-resolution images of Earth's surface, providing valuable data on deforestation, glacier retreat, and changes in land use. Instruments like the Hyperspectral Imager for the Coastal Ocean (HICO) and the Orbiting Carbon Observatory-3 (OCO-3) have been used to study the health of marine ecosystems and measure carbon dioxide levels in the atmosphere. These observations help scientists understand the impact of human activity on the environment and develop strategies for mitigating climate change.

The ISS also plays a crucial role in disaster management by providing real-time imagery of areas affected by natural disasters, such as hurricanes, wildfires, and floods. This information is used by emergency responders to assess the extent of damage, identify areas in need of assistance, and coordinate relief efforts. The station's unique perspective allows for a comprehensive view of large-scale events, providing data that can be used to improve disaster preparedness and response. The SERVIR program, a collaboration between NASA and the U.S. Agency for International Development (USAID), uses data from the ISS to support environmental decision-making and improve resilience in vulnerable communities around the world.

Atmospheric research is another area where the ISS has made significant contributions. The station's position above most of Earth's atmosphere allows scientists to study atmospheric phenomena without the interference that ground-based observations face. Instruments like the Stratospheric Aerosol and Gas Experiment (SAGE III) and the Lightning Imaging Sensor (LIS) have been used to study ozone concentrations, monitor aerosol levels, and observe lightning activity. These studies help scientists understand the complex processes that influence Earth's climate and weather patterns, contributing to more accurate climate models and improved weather forecasting.

In addition to observing Earth, the ISS is a valuable platform for space-based research. The Alpha Magnetic Spectrometer (AMS-02), a particle physics experiment installed on the station, is designed to study cosmic rays and search for evidence of dark matter and antimatter. The data collected by AMS-02 has provided insights into the composition of the universe and the fundamental forces that govern it. The ISS has also been used to study the behavior of cosmic dust, observe solar activity, and conduct astronomical observations. The station's ability to host a wide range of instruments makes it a versatile platform for advancing our understanding of both Earth and the cosmos.

International Research Collaborations

The International Space Station is a symbol of international cooperation, bringing together the efforts of space agencies from around the world to conduct groundbreaking research. The collaborative nature of the ISS extends to its scientific endeavors, with researchers from multiple countries contributing to experiments and sharing data. This spirit of cooperation has led to significant advancements in science and technology, demonstrating the value of working together to address complex challenges.

One of the key aspects of international collaboration on the ISS is the sharing of research facilities and resources. The station is equipped with laboratories from different space agencies, including NASA's Destiny Laboratory, ESA's Columbus Laboratory, and JAXA's Kibo Laboratory. These laboratories are open to researchers from all partner countries, allowing scientists to conduct experiments in a wide range of disciplines. The availability of shared resources has led to a diverse portfolio of research, with experiments covering everything from fluid dynamics to plant biology. By pooling resources and expertise, the ISS partners have been able to achieve more than any one nation could accomplish alone.

Joint research initiatives are another hallmark of international collaboration on the ISS. Many experiments conducted aboard the station involve partnerships between researchers from different countries, leveraging the unique capabilities of each partner to achieve common goals. For example, the Protein Crystal Growth experiments have involved scientists from the United States, Japan, and Europe, working together to study the crystallization of proteins in microgravity. The results of these experiments have implications for drug development, as understanding the structure of proteins can lead to the creation of more effective pharmaceuticals. The success of these joint initiatives highlights the benefits of international cooperation in advancing scientific knowledge.

The ISS has also played a role in fostering educational collaborations and inspiring the next generation of scientists and engineers. Space agencies from the ISS partner countries have developed educational programs that allow students to participate in research conducted aboard the station. Programs like ESA's Astro Pi Challenge and NASA's Student Spaceflight Experiments Program (SSEP) give students the opportunity to design and conduct experiments in space, providing a unique learning experience that encourages interest in science, technology, engineering, and mathematics (STEM). These educational initiatives are helping to build a global community of future scientists and engineers who are inspired by the possibilities of space exploration.

The international nature of the ISS has also led to the development of standardized protocols and procedures for conducting research in space. The diverse range of experiments conducted aboard the station requires careful coordination to ensure that resources are used efficiently and that the safety of the crew is maintained. The ISS partners have developed a set of standardized procedures for planning, executing, and analyzing experiments, which has helped streamline the research process and ensure that the results are reliable and reproducible. This standardization has not only benefited research conducted aboard the ISS but has also set a precedent for future international collaborations in space.

The collaborative research conducted aboard the ISS has led to a number of significant scientific breakthroughs that have benefited humanity. From advancements in medical research to the development of new technologies, the ISS has demonstrated the power of international cooperation in addressing complex challenges. The partnerships formed through the ISS have not only advanced our understanding of the universe but have also strengthened relationships between nations, fostering a spirit of unity and shared purpose. The scientific contributions of the ISS are a testament to what can be achieved when countries work together toward a common goal, and they serve as a model for future collaborations in space exploration and beyond.

The work conducted aboard the ISS represents the collective efforts of thousands of scientists, engineers, and astronauts from around the world. The research carried out on the station has advanced our understanding of human health, driven technological innovation, provided valuable insights into Earth's environment, and deepened our knowledge of the universe. The International Space Station stands as a testament to the power of international collaboration in advancing science and technology, and its contributions will continue to benefit humanity for years to come.

Chapter 6

Challenges and Crises

Technical Maintenance and Repairs

The International Space Station is a marvel of engineering, but maintaining it requires constant vigilance and effort. With hundreds of interconnected systems and components, the station is a complex network that needs regular maintenance to keep it functioning smoothly. Unlike Earth-based facilities, every repair on the ISS must consider the unique conditions of space, where gravity is absent, temperatures fluctuate dramatically, and access to replacement parts is limited. The crew must rely on their training and ingenuity, often becoming makeshift engineers when challenges arise.

One of the key aspects of ISS maintenance is ensuring that all life-support systems function without fail. The Environmental Control and Life Support System (ECLSS), which provides breathable air, clean water, and temperature control, requires regular checks and repairs to ensure the safety of the crew. Any failure in this system would endanger the lives of astronauts, so redundancies are built into the design. Astronauts are trained to conduct complex repairs, often working with guidance from engineers on the ground. These repairs can be time-consuming, requiring multiple spacewalks or extensive work inside the station to fix or replace malfunctioning parts.

Repairs in the vacuum of space pose unique challenges. Spacewalks, or extravehicular activities (EVAs), are often required to fix external systems, such as solar arrays, radiators, or communications equipment. These spacewalks are meticulously planned, with astronauts rehearsing the procedures in underwater facilities on Earth to simulate the microgravity environment. One notable repair involved a cooling system malfunction in 2010. The ammonia pump module, which is part of the thermal control system, failed and needed to be replaced. The repair required three spacewalks over several days, during which astronauts worked in shifts to remove the faulty pump and install a replacement. The complexity of the task, combined with the harsh environment of space, made it one of the most challenging repairs in ISS history.

Another critical area of maintenance is the power supply. The ISS is powered by large solar arrays, which convert sunlight into electricity. These arrays are mounted on trusses and need to be adjusted regularly to maximize energy collection. Over time, the arrays degrade due to exposure to space radiation and micrometeoroid impacts. Periodic replacements and adjustments are required to keep the station powered. In 2007, a torn solar panel necessitated an emergency repair. Astronaut Scott Parazynski conducted a spacewalk to repair the damage, using specially designed tools and clamps to stabilize the panel and restore its functionality. Such repairs highlight the ingenuity and adaptability required of the ISS crew to ensure uninterrupted power for the station's systems.

The Canadarm2, a robotic arm used for assembly and maintenance tasks, also requires regular upkeep. This robotic arm is crucial for capturing incoming cargo vehicles, installing new modules, and assisting astronauts during EVAs. Maintenance of the Canadarm2 includes replacing worn-out components, such as the Latching End Effectors (LEEs), which are the "hands" of the arm. These repairs often require a combination of remote operations from ground control and hands-on work by astronauts during spacewalks. The complexity of maintaining the robotic systems underscores the importance of both human and robotic collaboration in keeping the ISS operational.

Dealing with Space Hazards

The International Space Station operates in one of the most hazardous environments imaginable. Orbiting at an altitude of approximately 400 kilometers above Earth, the station faces a constant barrage of potential threats, including micrometeoroids, space debris, and radiation. The crew must remain vigilant, and the station's systems are designed to mitigate these risks as much as possible. Despite these precautions, dealing with space hazards is a constant challenge that requires careful planning and rapid response when issues arise.

Space debris is one of the most significant threats to the ISS. Traveling at speeds of up to 28,000 kilometers per hour, even small fragments of debris can cause serious damage if they collide with the station. The ISS is equipped with Whipple shielding, which consists of multiple layers designed to absorb and disperse the energy of impacts from smaller debris. For larger objects that pose a collision risk, NASA and other space agencies use radar to track debris and predict potential close approaches, known as conjunctions. When the risk of a collision is deemed significant, the station performs a Debris Avoidance Maneuver (DAM) to adjust its orbit and move out of the way. These maneuvers require careful coordination between ground control and the crew to ensure that all systems are prepared for the change in orbit.

Micrometeoroids, natural fragments of rock or metal from space, present another hazard. Unlike space debris, micrometeoroids are not tracked, making them unpredictable and potentially dangerous. The Whipple shielding on the station provides protection against these small, high-velocity particles, but impacts still occur. In 2018, a micrometeoroid strike caused a small pressure leak in the Russian Soyuz spacecraft docked to the ISS. The crew quickly located and sealed the leak using epoxy, preventing further pressure loss. Such incidents underscore the importance of rapid detection and response to ensure the safety of the crew.

Radiation is another significant hazard for the ISS. Without Earth's atmosphere to provide protection, astronauts are exposed to higher levels of cosmic radiation and solar particles. This radiation can increase the risk of cancer and cause other health issues over time. The ISS is equipped with radiation-hardened modules where the crew can take shelter during periods of increased solar activity, such as solar flares. Dosimeters worn by astronauts and installed throughout the station help monitor radiation levels, allowing mission control to track exposure and manage risk. Research conducted on the ISS has also contributed to the development of better radiation shielding materials, which will be crucial for future missions to the Moon and Mars.

The ISS crew undergoes extensive training to prepare for emergency situations related to space hazards. Drills are conducted regularly to practice responding to events such as rapid depressurization, fire, or toxic atmosphere. In the case of a major impact or other emergency, the crew can take refuge in the Soyuz or SpaceX Crew Dragon spacecraft, which are always docked to the station and ready to be used as lifeboats. These drills ensure that the crew is ready to respond swiftly and effectively to any hazard that may threaten the station's integrity or their safety.

The hazards faced by the ISS are a reminder of the challenges inherent in space exploration. The station's continued operation, despite these dangers, is a testament to the resilience of the systems designed to protect it and the training of the astronauts who live there. Every successful maneuver to avoid debris, every emergency repair, and every response to a radiation event is a demonstration of the commitment to maintaining a safe environment for the crew while pushing the boundaries of human presence in space.

Political and Budgetary Struggles

The International Space Station is not only a scientific and engineering achievement but also a political one. Its existence and continued operation depend on the collaboration of multiple countries, each with its own political and financial constraints. The complexities of international relations and the challenges of securing funding have been constant hurdles for the ISS program. Despite these struggles, the commitment of partner nations has allowed the station to remain operational and continue to make significant contributions to science and technology.

The ISS began as a vision of international cooperation, with the United States, Russia, Europe, Japan, and Canada coming together to create a permanent human presence in space. The political landscape at the time was marked by the end of the Cold War, which provided an opportunity for former adversaries to collaborate on a peaceful project. However, maintaining this partnership has not always been easy. Changes in government leadership, shifting national priorities, and economic challenges have all impacted the funding and support for the ISS over the years.

In the United States, the ISS has faced numerous budgetary challenges. The high cost of building and maintaining the station has led to debates in Congress about the value of the program. At times, funding for the ISS has been threatened by budget cuts or shifts in focus toward other space exploration goals, such as returning to the Moon or sending humans to Mars. NASA has had to make a strong case for the benefits of the ISS, emphasizing its role in scientific research, international cooperation, and as a stepping stone for future exploration. The support of international partners has also been crucial in convincing U.S. lawmakers of the importance of maintaining the ISS.

Russia, a key partner in the ISS, has also faced significant challenges in maintaining its commitment to the station. Economic instability, particularly following the collapse of the Soviet Union, made it difficult for Russia to contribute financially to the program. Despite these challenges, Russia has remained a vital partner, providing essential modules, crew transportation, and technical expertise. The partnership between NASA and Roscosmos has been tested at times by geopolitical tensions, but both agencies have continued to work together to ensure the success of the ISS. The cooperation between the United States and Russia on the ISS is often cited as an example of how space exploration can bridge political divides and foster collaboration, even during times of international conflict.

The European Space Agency (ESA), Japan's JAXA, and the Canadian Space Agency (CSA) have also faced budgetary pressures that have impacted their contributions to the ISS. Each of these agencies must secure funding from their respective governments, which can be challenging given competing priorities and economic constraints. Despite these challenges, ESA, JAXA, and CSA have remained committed partners, providing important modules, scientific experiments, and technical support. The success of the ISS as an international partnership is a testament to the willingness of these nations to invest in a shared vision of space exploration and scientific discovery.

One of the most significant political challenges facing the ISS is the question of its future. The station was originally intended to operate until 2020, but its lifespan has since been extended to 2030. As the station ages, the costs of maintaining and upgrading its systems will continue to rise, leading to questions about the long-term sustainability of the program. Discussions are ongoing among the ISS partners about the future of the station, including the possibility of transitioning to a commercially operated platform or developing new international partnerships for future space stations. The political and financial decisions made in the coming years will determine the future of human presence in low Earth orbit and the direction of international space exploration efforts.

The political and budgetary struggles faced by the ISS highlight the challenges of maintaining an international project of this scale. Despite these challenges, the commitment of partner nations to the vision of a permanent human presence in space has allowed the station to thrive. The ISS serves as a model for international cooperation, demonstrating that, despite political differences and financial constraints, nations can work together to achieve something greater than any one country could accomplish alone.

Crisis Management on Board

Life aboard the International Space Station is inherently risky, and the crew must be prepared to handle a wide range of emergencies.

Crisis management is a critical aspect of living and working in space, where even minor issues can escalate into life-threatening situations. The crew is extensively trained to deal with emergencies such as fire, rapid depressurization, toxic atmosphere, and medical crises. The ability to respond quickly and effectively to these situations is crucial for ensuring the safety of the astronauts and the continued operation of the station.

One of the most serious emergencies that can occur on the ISS is a fire. Fire in space is particularly dangerous because it can spread rapidly in a confined environment, and the behavior of flames in microgravity is different from that on Earth. Fire tends to form spherical shapes, and the lack of gravity affects the way heat and smoke move. The ISS is equipped with smoke detectors, fire extinguishers, and portable breathing gear to help the crew respond to a fire. Regular fire drills are conducted to ensure that all crew members are familiar with emergency procedures, including the use of fire extinguishers and the steps to take to isolate and extinguish a fire. The crew must also be prepared to quickly don breathing masks and move to a safe area if smoke or toxic gases are present.

Rapid depressurization is another critical emergency scenario that the crew must be prepared for. A breach in the station's hull, whether caused by a micrometeoroid impact or a mechanical failure, can lead to a rapid loss of air pressure. The ISS is designed with multiple layers of shielding to protect against impacts, and the station's modules can be sealed off from one another in the event of a breach. The crew is trained to quickly locate and seal leaks using specialized equipment, such as patch kits and sealant. Drills are conducted regularly to practice responding to a depressurization event, with crew members rehearsing the steps to locate the source of the leak, isolate the affected module, and restore pressure. The ability to respond swiftly to a depressurization event is essential for maintaining the safety of the crew.

Toxic atmosphere is another potential crisis that the crew must be prepared to handle. The ISS is equipped with sensors that monitor the air quality, detecting harmful gases such as ammonia or carbon dioxide. If a leak or malfunction leads to the release of a toxic substance, the crew must act quickly to identify the source and take corrective action. This may involve isolating a specific module, using portable air scrubbers to remove contaminants, or donning protective gear. The crew is trained to work together to assess the situation, communicate with mission control, and implement the necessary measures to restore a safe environment.

Medical emergencies are also a significant concern aboard the ISS. The crew has access to a range of medical supplies and equipment, and all astronauts receive medical training before their mission. This training includes basic first aid, the use of medical instruments, and the ability to perform minor surgical procedures if necessary. Telemedicine plays a crucial role in medical care on the ISS, allowing astronauts to consult with doctors on Earth in real time. In the event of a serious medical emergency, the crew has the option to evacuate the affected astronaut to Earth aboard the Soyuz or SpaceX Crew Dragon spacecraft. The ability to respond effectively to medical emergencies is critical for maintaining the health and well-being of the crew during their time in space.

The crew's ability to manage crises on board the ISS is a testament to their extensive training and the careful design of the station's systems. Every astronaut undergoes rigorous preparation for dealing with emergencies, including simulations that replicate potential scenarios they may face in space. These simulations help the crew develop the skills and confidence needed to respond effectively to unexpected situations. The coordination between the crew on board and mission control on the ground is also essential for managing crises, as ground-based experts provide guidance and support during emergencies.

Crisis management on the ISS is an ongoing process, with regular drills, equipment checks, and system updates to ensure that the station remains a safe environment for the crew. The ability to respond effectively to emergencies is what allows the ISS to continue its mission of scientific research and international cooperation, despite the inherent risks of living and working in space. The experiences gained from managing crises on the ISS are also helping to inform the development of future space habitats, ensuring that lessons learned in low Earth orbit are applied to missions beyond our planet.

Chapter 7

Legacy and Future

A Symbol of Peaceful Collaboration

The International Space Station stands as one of the most significant symbols of peaceful collaboration in human history. Unlike earlier space endeavors, which were often motivated by competition, particularly during the Cold War era, the ISS was conceived as a platform for unity. Bringing together the space agencies of the United States, Russia, Europe, Japan, and Canada, the ISS represents what humanity can achieve when nations work towards a common goal, transcending political tensions and national borders. This collaborative spirit is particularly striking considering the history of space exploration, which was once marked by intense rivalry, especially between the United States and the Soviet Union. The partnership that led to the ISS began with a shared vision of creating a permanent human outpost in low Earth orbit. This vision required the pooling of resources, expertise, and political will from countries that had previously viewed each other as adversaries. The United States and Russia, two of the primary partners, exemplified this shift from competition to collaboration. During the Cold War, these two nations were engaged in a fierce space race, each striving to outdo the other in technological achievements. But with the fall of the Soviet Union and the end of the Cold War, the opportunity arose for a new era of cooperation. The ISS became a vehicle for demonstrating that even former adversaries could work together to achieve a greater good.

The international nature of the ISS extends beyond the participation of national space agencies. It includes a diverse array of scientists, engineers, and astronauts from around the world, each contributing their skills and knowledge to the station's mission. The station has hosted astronauts from a wide range of countries, including many that are not part of the core partnership. This inclusivity has made the ISS a symbol of global unity, demonstrating that space exploration is not the domain of a select few nations but a shared endeavor for all of humanity. The experiences of astronauts from different cultural backgrounds living and working together in space have provided valuable lessons in teamwork, communication, and the importance of diverse perspectives.

The ISS has also played a role in fostering peaceful collaboration through its contributions to international treaties and cooperative frameworks. The Intergovernmental Agreement (IGA), which governs the operation of the ISS, is a model for how nations can work together in space. This agreement outlines the roles and responsibilities of each partner, ensuring that all contributions are recognized and that the benefits of the station are shared. The success of the ISS partnership has inspired similar international collaborations, such as the Artemis Accords, which aim to establish principles for the peaceful exploration of the Moon and beyond.

The legacy of the ISS as a symbol of peaceful collaboration is one that will endure long after the station itself is retired. It has shown that when nations set aside their differences and work together, they can achieve remarkable things. The lessons learned from the ISS partnership will continue to inform future international efforts in space exploration, providing a foundation for cooperation as humanity sets its sights on more distant destinations, such as the Moon, Mars, and beyond.

The End of an Era: Future Plans

The International Space Station has been in continuous operation for over two decades, but its time in orbit is not indefinite. As the station ages, discussions about its future have become increasingly important. Originally designed to operate until 2020, the ISS's mission has been extended multiple times, with the current plan to continue operations until at least 2030. Beyond that, the future of the ISS will depend on a combination of technical, financial, and political factors, as well as the evolving goals of the international space community.

One of the key considerations for the future of the ISS is the increasing cost of maintenance and the challenges associated with keeping the aging station operational. Many of the station's components have been in orbit since the late 1990s and early 2000s, and while regular maintenance and upgrades have kept the station functional, the cost of these efforts continues to rise. As the station ages, the risk of component failures increases, necessitating more frequent repairs and replacements. The financial burden of maintaining the ISS falls on the partner space agencies, and there is growing interest in finding ways to reduce these costs, either by transitioning to a new model of operation or by gradually phasing out the station.

One potential path for the future of the ISS involves transitioning to commercial operations. NASA and other partner agencies have expressed interest in opening up the station to greater commercial involvement, allowing private companies to use the ISS for research, manufacturing, and even tourism. This shift would help offset the costs of maintaining the station and could serve as a stepping stone for the development of commercially operated space stations. Companies like Axiom Space are already planning to attach commercial modules to the ISS, with the goal of eventually operating independent space stations once the ISS is retired. This transition represents a shift in how space infrastructure is managed, with a greater emphasis on public-private partnerships and the commercialization of low Earth orbit.

Another aspect of the future of the ISS is its role in preparing for deep space exploration. The ISS has provided invaluable experience in long-duration spaceflight, life support, and international collaboration—all of which are critical for future missions to the Moon and Mars. As NASA and its partners work towards the Artemis program, which aims to return humans to the Moon, the ISS will continue to serve as a testbed for technologies and systems that will be used in deep space. The knowledge gained from operating the ISS will help ensure the success of future missions beyond low Earth orbit, making the station an important part of the broader strategy for human exploration of the solar system.

Eventually, the ISS will reach the end of its operational life, and plans will need to be made for its deorbiting and safe disposal. This process will be carefully managed to ensure that the station reenters Earth's atmosphere in a controlled manner, with any remaining debris falling into an uninhabited area of the ocean, often referred to as the "spacecraft cemetery." The deorbiting of the ISS will mark the end of an era in human spaceflight, but it will also pave the way for new opportunities in space exploration. Whether through the development of new international space stations, commercially operated platforms, or missions to the Moon and Mars, the legacy of the ISS will live on in the next generation of space endeavors.

Preparing for Deep Space Exploration

The International Space Station has played a crucial role in preparing humanity for deep space exploration. The lessons learned from living and working on the ISS are helping to lay the groundwork for future missions to the Moon, Mars, and beyond. The unique environment of the ISS has provided an opportunity to test the technologies, systems, and human factors that will be essential for the success of long-duration missions in deep space.

One of the most important contributions of the ISS to deep space exploration is the development of life support systems capable of sustaining human life for extended periods. The Environmental Control and Life Support System (ECLSS) on the ISS has been a key area of research, providing valuable insights into the recycling of air and water, waste management, and the maintenance of a habitable environment in space. These systems are being refined and improved based on the experience gained on the ISS, with the goal of developing more efficient and reliable life support systems for missions to the Moon and Mars. The ability to recycle resources and minimize the need for resupply will be critical for deep space missions, where resupply from Earth will be limited or impossible.

The ISS has also been instrumental in studying the effects of long-duration spaceflight on the human body. Prolonged exposure to microgravity presents a number of challenges, including muscle atrophy, bone density loss, fluid shifts, and changes in the cardiovascular and immune systems. Understanding these effects and developing countermeasures has been a major focus of research on the ISS. The data collected from astronauts who have spent months, and in some cases over a year, on the ISS is helping scientists understand how the human body adapts to space and what measures can be taken to mitigate the negative effects of microgravity. This research is essential for ensuring the health and well-being of astronauts on future missions to Mars, which could last for more than two years.

Radiation exposure is another major challenge for deep space exploration, and the ISS has provided valuable insights into how to manage this risk. Unlike low Earth orbit, where the Earth's magnetic field provides some protection from cosmic radiation, deep space missions will expose astronauts to much higher levels of radiation. The ISS has allowed researchers to study the effects of radiation on both astronauts and equipment, helping to develop better shielding materials and strategies for minimizing exposure. Understanding how to protect astronauts from radiation will be critical for the success of missions beyond low Earth orbit, where the risks are significantly higher.

The experience of operating the ISS has also provided valuable lessons in spacecraft design, maintenance, and logistics. The challenges of maintaining and repairing the ISS have highlighted the importance of designing systems that are modular, redundant, and easy to service. These principles are being applied to the design of future spacecraft, such as NASA's Orion spacecraft and the Lunar Gateway, which will serve as a staging point for missions to the lunar surface. The Gateway, in particular, will build on the lessons learned from the ISS, providing a platform for long-duration habitation and serving as a testbed for technologies that will be used on Mars missions.

The ISS has also been a proving ground for international collaboration in space exploration. The partnerships formed through the ISS program have demonstrated the value of working together to achieve common goals, and these partnerships will be critical for future deep space missions. The Artemis program, which aims to return humans to the Moon and eventually send them to Mars, is being developed as an international effort, with contributions from NASA, ESA, JAXA, and other space agencies. The experience of working together on the ISS has helped build the relationships and trust needed for these future missions, ensuring that humanity's exploration of deep space will be a collaborative endeavor.

Inspiring Future Generations

The International Space Station has not only advanced science and technology but has also served as a powerful source of inspiration for people around the world. The sight of astronauts floating weightlessly, conducting experiments, and gazing down at Earth from the Cupola has captured the imaginations of millions. The ISS has played a significant role in inspiring future generations to pursue careers in science, technology, engineering, and mathematics (STEM), helping to ensure that the next wave of explorers, scientists, and engineers is ready to take on the challenges of space exploration.

One of the most impactful ways the ISS has inspired future generations is through educational outreach. Space agencies like NASA, ESA, and JAXA have developed a wide range of educational programs that allow students to engage with the science and technology of the ISS. Programs such as NASA's STEM on Station and ESA's Astro Pi Challenge give students the opportunity to design and conduct experiments that are carried out aboard the ISS. These programs provide a unique hands-on learning experience that helps students understand the principles of science and engineering while sparking their interest in space exploration.

Astronauts aboard the ISS have also played a key role in inspiring young people by sharing their experiences through live video calls, social media, and educational broadcasts. The ability to communicate directly with astronauts in space has made the ISS more accessible to the public and has provided a personal connection to the work being done in orbit. Astronauts like Chris Hadfield, who famously performed a cover of David Bowie's "Space Oddity" from the ISS, have used social media to share their experiences in space with millions of followers. These efforts have helped demystify space exploration, making it relatable and inspiring people of all ages to dream about the possibilities of human spaceflight.

The ISS has also served as a platform for international cultural exchange, highlighting the diversity of the human experience. The crew members aboard the ISS come from different countries and backgrounds, and their ability to live and work together in space is a powerful symbol of what humanity can achieve when we embrace our differences and work towards a common goal. The station's international nature has inspired young people around the world to see themselves as part of a global community, united by a shared curiosity and a desire to explore the unknown. The ISS shows that space exploration is not limited to one nation or one group of people—it is a collective endeavor that benefits all of humanity.

The ISS has also been a source of inspiration for artists, writers, and filmmakers. The imagery of the station orbiting above Earth, the views of our planet from space, and the stories of the astronauts who live there have all influenced popular culture. Films like "Gravity" and "The Martian" have drawn on the experiences of astronauts aboard the ISS to create realistic depictions of life in space, capturing both the challenges and the wonder of space exploration. These cultural representations have helped to keep the public engaged with the ISS and have inspired a new generation of storytellers to imagine the future of human spaceflight.

The legacy of the ISS as a source of inspiration will continue long after the station itself is retired. The lessons learned from the ISS, the stories of the astronauts who lived and worked there, and the scientific discoveries made in orbit will all serve as a foundation for the next generation of explorers. The ISS has shown that space exploration is not just about advancing technology—it is about pushing the boundaries of what is possible, inspiring people to look beyond their immediate surroundings, and encouraging them to reach for the stars. The young people who have been inspired by the ISS today will be the ones who design, build, and crew the missions to the Moon, Mars, and beyond in the years to come.

The impact of the ISS on future generations is perhaps its greatest legacy. By showing what is possible when nations work together, by advancing our understanding of science and technology, and by inspiring millions of people around the world, the ISS has paved the way for the future of human space exploration. The spirit of curiosity, collaboration, and exploration that defines the ISS will continue to inspire humanity for generations to come, ensuring that the dream of space exploration lives on in the hearts and minds of people everywhere.

Appendices

A: Timeline

Early Concepts (1950s-1970s):

> Wernher von Braun's Vision: The concept of a space station dates back to the 1950s when Wernher von Braun, a pioneer of rocket technology, proposed an ambitious plan for a large, wheel-shaped space station. His vision, which included a rotating structure to create artificial gravity, influenced many subsequent ideas about space stations and space exploration.
>
> NASA's Space Station Concepts: Throughout the 1960s and 1970s, NASA explored multiple space station designs, including the Manned Orbiting Laboratory (MOL) and Skylab. Skylab, launched in 1973, was the United States' first space station and served as a key precursor to the ISS, providing valuable insights into long-duration human spaceflight.
>
> Interkosmos and Salyut Programs: In parallel, the Soviet Union launched a series of space stations under the Salyut and later the Mir programs. Mir, in particular, served as a major milestone in space station technology and international collaboration, as it hosted astronauts from various countries.

1984: President Ronald Reagan announces the U.S. commitment to build an international space station, inviting other nations to participate.

1993: Russia joins the ISS program, marking the beginning of a major international partnership.

1998: The first module, Zarya, is launched into orbit, followed shortly by the Unity module, marking the start of ISS assembly.

2000: The Zvezda Service Module is launched, providing essential life support, and the first crew, Expedition 1, arrives at the ISS.

2001: Destiny Laboratory is added, allowing more extensive scientific research to begin.

2008: The European Space Agency's Columbus Laboratory and Japan's Kibo Laboratory are added, significantly expanding the station's research capabilities.

2011: The Space Shuttle is retired, ending NASA's primary means of transporting astronauts and heavy cargo to the ISS.

2019: NASA announces the opening of the ISS to commercial use, including space tourism and research by private companies.

2020: SpaceX's Crew Dragon spacecraft successfully carries astronauts to the ISS, marking the return of human spaceflight capability to the United States.

2030 (Projected): Planned retirement of the ISS, with discussions of transitioning to commercial platforms and new international initiatives.

B: ISS Partners and Contributions

1. NASA (United States)
    - Key Contributions: Coordination and overall leadership, Destiny Laboratory, Truss segments, solar arrays, logistics support.
    - Significant Modules: Unity, Harmony, Tranquility, Destiny.
2. Roscosmos (Russia)
    - Key Contributions: Propulsion modules, life-support systems, crew transportation.

- Significant Modules: Zarya, Zvezda, docking ports, Soyuz and Progress spacecraft.
3. European Space Agency (ESA)
    - Key Contributions: Scientific research, logistics support, collaboration in astronaut training.
    - Significant Modules: Columbus Laboratory, Automated Transfer Vehicle (ATV).
4. Japan Aerospace Exploration Agency (JAXA)
    - Key Contributions: Research capabilities, logistics support, experiments related to biology and material sciences.
    - Significant Modules: Kibo Laboratory, H-II Transfer Vehicle (HTV).
5. Canadian Space Agency (CSA)
    - Key Contributions: Robotics systems for assembly and maintenance.
    - Significant Contribution: Canadarm2, Dextre (robotic manipulator).
6. Italian Space Agency (ASI)
    - Key Contributions: Provided pressurized cargo modules and support in research.
    - Significant Contribution: Multipurpose Logistics Module (MPLM).
7. Brazilian Space Agency (AEB)
    - Key Contributions: Provided scientific equipment and experiments.
    - Significant Contribution: Participation in research collaborations.
8. Centre National d'Études Spatiales (CNES) - France
    - Key Contributions: Contributed to research and provided technology used in experiments aboard the ISS.
9. German Aerospace Center (DLR)

- Key Contributions: Assisted in developing scientific payloads and research in life sciences and material sciences.
- Significant Contribution: Experiments in material processing and biological studies.

10. United Arab Emirates Space Agency (UAESA)
    - Key Contributions: Sponsored astronaut missions to the ISS and contributed to scientific research.
    - Significant Contribution: Emirati astronaut missions to the ISS for international collaboration.
11. Indian Space Research Organisation (ISRO)
    - Key Contributions: Provided research contributions and experimental payloads, and collaborated in areas such as life sciences and materials research.
    - Significant Contribution: Participation in space science experiments and providing satellite data to support ISS operations.

C: Life-Support Systems on the ISS

- Environmental Control and Life Support System (ECLSS): Provides clean air and water. Recycles water from urine and humidity, creating up to 90% of potable water used on the ISS.

- Carbon Dioxide Removal Assembly (CDRA): Filters and removes excess CO_2 from the station's atmosphere to maintain safe levels.

- Oxygen Generation Assembly (OGA): Electrolyzes water to produce oxygen for the crew to breathe.

- Temperature Control: Maintains a stable internal temperature using heat exchangers and radiators to dissipate excess heat.

D: Significant ISS Experiments

1. Microgravity Combustion Studies: Improved understanding of flame behavior, leading to safer designs for spacecraft and fire safety advancements on Earth.

2. Protein Crystal Growth: Enabled the production of high-quality protein crystals, aiding in the development of new pharmaceuticals.

3. Advanced Plant Habitat: Research on plant growth in microgravity has contributed to our understanding of sustainable food production techniques that can be used in space and on Earth.

4. Fluid Dynamics: Experiments have provided insights into how fluids behave in microgravity, informing the design of more efficient fuel systems for rockets and improvements in medical devices such as syringes.

E: Major ISS Spacewalks

- STS-88 (1998): The first spacewalk to connect the Zarya and Unity modules, marking the beginning of ISS assembly.

- Ammonia Pump Replacement (2010): Multiple spacewalks to replace a failed ammonia pump, restoring the cooling system's functionality.

- Solar Array Repair (2007): Emergency spacewalk conducted by astronaut Scott Parazynski to repair a torn solar array, ensuring the station's power generation capacity.

- Alpha Magnetic Spectrometer Installation (2011): Installation of the AMS-02, a major particle physics experiment aimed at studying cosmic rays.

F: ISS Educational Initiatives

- Astro Pi (ESA): European students participate in coding competitions, with their code executed on Raspberry Pi computers aboard the ISS.

- NASA's Student Spaceflight Experiments Program (SSEP): Allows students from across the United States to design experiments that are conducted by astronauts on the ISS.

- Mission Discovery (UK): A program where young people work with astronauts and scientists to develop experiments that may be selected to fly to the ISS.

G: Key Technologies Developed

1. Water Recycling Systems: Efficient recycling of wastewater has led to improved filtration technologies for areas on Earth with limited access to clean water.

2. Robotics: The Canadarm2 and Dextre have influenced the development of robotics used in surgery and other precision applications on Earth.

3. Advanced Alloys and Materials: Studies of material properties in microgravity have contributed to the development of stronger and lighter alloys used in aerospace and manufacturing.

4. Telemedicine: Techniques developed to provide medical care to astronauts remotely have been adapted for use in rural and underserved areas on Earth.

Appendix H: Glossary of Key Terms

- Microgravity: The condition in which objects appear to be weightless, typically experienced in orbit.
- Extravehicular Activity (EVA): A spacewalk conducted by an astronaut outside the spacecraft.
- Soyuz: A Russian spacecraft used for transporting crew and cargo to the ISS.
- Lunar Gateway: A planned space station that will orbit the Moon, serving as a staging point for missions to the lunar surface and beyond.
- Alpha Magnetic Spectrometer (AMS-02): A particle physics experiment mounted on the ISS to study cosmic rays and search for dark matter.

I. The ISS FAQ

1. How Does the ISS Maintain Its Orbit?
- **Gravity and Velocity**: The ISS remains in orbit due to a balance between Earth's gravitational pull and its high velocity of about 28,000 kilometers per hour (17,500 miles per hour). This combination creates a state of perpetual free fall around Earth.
- **Orbital Mechanics**: The station follows a curved path, continuously falling towards Earth but moving forward fast enough to keep missing it, resulting in a stable orbit.
- **Thrusters and Gyroscopes**: Onboard thrusters and gyroscopes help maintain orientation and make minor altitude adjustments as needed.

2. Can the ISS Halt or Slow Down Over an Area?

- **Orbit Adjustments**: The ISS cannot come to a full stop in its orbit, as that would cause it to fall back to Earth. However, it can make small changes to its speed and altitude using thrusters.
- **Station-Keeping Maneuvers**: Minor adjustments can be made to avoid space debris or prepare for docking, but the ISS cannot hover over a specific point like a helicopter.

3. How Do Astronauts Sleep on the ISS?
- **Sleeping Compartments**: Each astronaut has a small sleeping compartment equipped with a sleeping bag. They strap themselves in to prevent floating around while sleeping.
- **No Up or Down**: In microgravity, there is no up or down, so astronauts can sleep in any orientation.

4. How Do Astronauts Stay Fit in Microgravity?
- **Daily Exercise**: Astronauts exercise for about two hours each day to counteract muscle atrophy and bone loss. They use treadmills, stationary bikes, and resistive exercise devices.
- **Equipment**: Specialized machines such as the Advanced Resistive Exercise Device (ARED) simulate weightlifting in space.

5. How Is Waste Managed on the ISS?
- **Solid Waste**: Solid waste is collected in containers and stored until it can be disposed of in cargo spacecraft that burn up upon reentry.
- **Liquid Waste**: Urine is processed by the Environmental Control and Life Support System (ECLSS) to be recycled into potable water.

6. How Does the ISS Get Oxygen?
- **Oxygen Generation Assembly (OGA)**: Water is split into hydrogen and oxygen through electrolysis, with the oxygen used for breathing.
- **Backup Supplies**: Oxygen can also be provided by tanks brought by cargo spacecraft or by burning chemical oxygen generators in emergencies.

7. How Is Water Recycled on the ISS?
- **Water Recovery**: The ECLSS recycles up to 90% of the water, including urine and condensation, into drinkable water.
- **Closed-Loop System**: The recycling process reduces the need for frequent resupply missions, conserving resources.

8. How Do Astronauts Deal With Space Debris?
- **Tracking and Avoidance**: NASA and other space agencies track space debris and can initiate Debris Avoidance Maneuvers (DAM) to move the ISS out of harm's way.
- **Whipple Shielding**: The station is equipped with shielding to protect against small debris impacts.

9. How Do They Handle Medical Emergencies?
- **Medical Training**: All astronauts receive medical training, including first aid and minor surgical procedures.
- **Telemedicine**: Crew members consult with doctors on Earth via real-time video communication.
- **Emergency Return**: In serious cases, astronauts can return to Earth quickly using the Soyuz or SpaceX Crew Dragon spacecraft.

10. Why Doesn't the ISS Have Artificial Gravity?

THE INTERNATIONAL SPACE STATION

- **Engineering Challenges**: Creating artificial gravity would require a rotating structure, adding significant complexity and weight.
- **Focus on Microgravity Research**: The ISS is designed for microgravity research, which is one of its primary missions.

11. How Is Electricity Generated on the ISS?
- **Solar Panels**: Large solar arrays convert sunlight into electricity. The panels rotate to track the Sun, optimizing energy collection.
- **Battery Storage**: Excess energy is stored in batteries for use when the ISS is in Earth's shadow.

12. How Often Do Astronauts Eat, and What Do They Eat?
- **Meals Per Day**: Astronauts eat three main meals per day, plus snacks. Food is typically pre-packaged and dehydrated for easy storage.
- **Variety**: Meals include a mix of dehydrated, thermostabilized, and vacuum-sealed foods. Special occasions are sometimes marked with special treats.

13. How Do Astronauts Clean Themselves Without a Shower?
- **No Shower Available**: There is no shower on the ISS. Astronauts use rinseless body wipes and no-rinse shampoo to stay clean.
- **Hand Hygiene**: Wet towels and soap are used for cleaning hands and faces.

14. Can the ISS Be Seen From Earth?
- **Visibility**: Yes, the ISS is visible to the naked eye as a bright, fast-moving object. It reflects sunlight, making it easy to spot at dawn or dusk.

- **Tracking Tools**: Various websites and apps provide tracking information, letting people know when and where to look for the ISS.

15. How Is Fire Safety Managed on the ISS?
- **Fire Detection**: Smoke detectors are installed throughout the station.
- **Fire Suppression**: Fire extinguishers and portable breathing equipment are available, and astronauts are trained in fire response procedures.

16. What Happens If a Module Develops a Leak?
- **Locating the Leak**: Pressure sensors help detect a drop in cabin pressure, and the crew uses ultrasonic leak detectors to locate it.
- **Sealing the Leak**: Leak repair kits, including epoxy and seal patches, are used to seal any leaks.

17. How Do Astronauts Communicate With Earth?
- **Radio Signals**: The ISS uses high-frequency radio signals and relay satellites to maintain continuous communication with mission control.
- **Video and Audio Links**: Real-time video and audio links allow astronauts to communicate with engineers, scientists, and even their families.

18. How Long Does It Take to Get to the ISS?
- **Travel Time**: It takes around six hours to two days to reach the ISS from Earth, depending on the mission profile and spacecraft used.

19. Why Does the ISS Need to Be Reboosted?

- **Orbital Decay**: The ISS experiences drag from Earth's atmosphere, causing it to gradually lose altitude. Periodic reboosts by thrusters or visiting spacecraft are needed to maintain its orbit.

20. What Is the Average Duration of an Astronaut's Stay on the ISS?

- **Typical Mission Length**: Astronauts usually stay for about six months, although some missions have lasted up to a year to study the effects of long-duration spaceflight.

21. Can Astronauts Vote From Space?

- **Yes**: U.S. astronauts can vote while aboard the ISS through a secure electronic absentee ballot, a process set up in coordination with NASA and local election officials.

22. How Does the ISS Handle Trash?

- **Trash Disposal**: Trash is packed into cargo ships, like the Cygnus or Progress, which burn up upon reentry into Earth's atmosphere, effectively incinerating the waste.

23. How Do They Deal With Time Zones on the ISS?

- **Coordinated Universal Time (UTC)**: The ISS follows UTC to maintain a consistent schedule. This helps synchronize activities with mission control centers around the world.

24. How Do Astronauts Stay Mentally Healthy?

- **Leisure Activities**: Crew members have scheduled free time for hobbies, such as reading, watching movies, or looking out of the Cupola.
- **Communication With Family**: Regular video calls and emails help astronauts stay connected to their families.

25. What Is the Cupola?
- **Observation Module**: The Cupola is a module with seven windows that provides panoramic views of Earth. It is often used for observation, photography, and relaxation.

26. How Are Spacewalks Planned and Conducted?
- **Planning**: Spacewalks, or EVAs, are meticulously planned months in advance to address maintenance needs or conduct installations.
- **Execution**: Astronauts rehearse procedures in underwater training facilities to simulate microgravity, and they use safety tethers and specialized tools during the actual spacewalk.

27. How Is Food Stored and Prepared on the ISS?
- **Storage**: Food is stored in vacuum-sealed pouches to keep it fresh and prevent spoilage.
- **Preparation**: Astronauts use hot water dispensers and food warmers to rehydrate and heat their meals. There is no conventional oven or stove.

28. How Do Astronauts Deal With the Lack of Fresh Food?
- **Fresh Food Deliveries**: Cargo resupply missions occasionally bring fresh fruits and vegetables, which are eaten quickly before they spoil.
- **Supplements**: The diet is supplemented with dehydrated foods, and astronauts also grow small amounts of fresh produce, like lettuce, in the Veggie experiment.

29. How Are Tools Prevented From Floating Away During Repairs?

- **Tool Tethers**: All tools used during spacewalks and internal repairs are tethered to the astronaut or the workstation.
- **Tool Belts and Bags**: Astronauts wear tool belts or use specially designed bags to keep their equipment secure and organized.

30. How Do Astronauts Deal With Sudden Changes in Altitude or Position?

- **Gyroscopes and Thrusters**: The ISS is equipped with Control Moment Gyroscopes (CMGs) and thrusters that help maintain its orientation and adjust its altitude when needed. These systems work together to manage the station's attitude, keeping it stable during various operations.
- **Crew Coordination**: Astronauts work closely with mission control to execute these adjustments safely, ensuring the station remains in the correct orbit and position for ongoing activities and experiments.

DR. LEO LEXICON

Thank you for reading this book. Please write a review, and share with your friends on social media if you enjoyed this title.
We are counting on you to spread the word!

Explore more titles from Lexicon Labs in the pages that follow.

Don't forget to sign up to our newsletter and download your FREE poster print! Go to https://mindzen.squarespace.com/ and sign up today!

Explore the lives of great innovators, Scientists, Leaders, Artists and Explorers...Stay tuned for additional titles coming soon!

*Learn the basics of Coding and program in Python.
No prior knowledge required!*

Meet our bestselling titles on AI
BOOKS FOR CURIOUS MINDS

- Structured introduction to the building blocks of AI
- Review of major milestones in AI history
- Meet the leading inventors and their key innovations
- AI concepts explained in a simple, easy-to-understand format by a Bay Area educator
- Resources for puzzles, games, and coding
- Perfect travel companion or gift

Follow Dr. Leo Lexicon on Twitter/X

 @LeoLexicon

LEARN ALL ABOUT STARTING AND GROWING A BUSINESS AS A TEENAGER

Explore the Future of Quantum Computing

FUN BOOKS FOR TRIVIA NIGHT

COLORING BOOKS

TEST YOUR INNER NERD!

 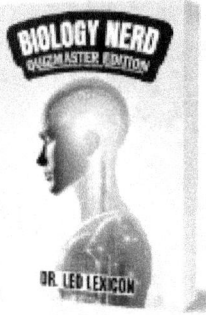

Discover More Bestselling Titles from Lexicon Labs!

SCAN ME

Education, Entertainment, and Inspiration. GUARANTEED.